Asceticism and Healing in Ancient India

Asceticism and Healing in Ancient India

Medicine in the Buddhist Monastery

KENNETH G. ZYSK

New York Oxford
OXFORD UNIVERSITY PRESS
1991

Oxford University Press

Oxford New York Toronto
Delhi Bombay Calcutta Madras Karachi
Petaling Jaya Singapore Hong Kong Tokyo
Nairobi Dar es Salaam Cape Town
Melbourne Auckland

and associated companies in
Berlin Ibadan

Published by Oxford University Press, Inc.,
200 Madison Avenue, New York, NY 10016

Library of Congress Cataloging-in-Publication Data

Zysk, Kenneth G.
Asceticism and healing in ancient India: medicine in the Buddhist monastery / Kenneth G. Zysk.
p. cm. Includes bibliographical references.
ISBN 0-19-505956-5
1. Medicine, Buddhist. 2. Medicine, Ayurvedic. 3. Medicine—Religious aspects—Buddhism. 4. Monastic and religious life (Buddhism) I. Title.
R605.Z87 1990 610′.934—dc20 89-23067

2 4 6 8 9 7 5 3 1

Printed in the United States of America
on acid-free paper

To the memory of my father, Stanley A. Zysk,
whose recent death stole away
even his anticipation of enjoying the fruits
of his constant encouragement and support

Preface

Human susceptibility to illness and injury suggests and extensive reading of Indian literature confirms that medical lore touched the lives of almost all Indians from Vedic Āryans to moderns in cosmopolitan or traditional settings. Ancient treatises, Hindu and non-Hindu alike, contain numerous metaphors, similes, and references to disease and healing. Although Hinduism tended to emphasize spiritual and ultimate reality, Indians throughout the centuries remained acutely aware of the physical factors that affected their existence and cut short their period of life on earth. Efforts in ancient India to subjugate, control, and understand these phenomena in order to mitigate their harmful effects and prolong one's earthly existence gave rise to a long tradition of healing arts that found expression in many types of religious and secular literature.

My own exploration of ancient Indian medical lore starts from its earliest beginnings and traces its development through the centuries in order to discover the roots of India's traditional system of medicine, *āyurveda* (the science of longevity). The following pages, which present the results of this investigation, offer a picture of ancient Indian medical history radically different from the one commonly portrayed. The sources show that the Hindu śāstric tradition of medicine derived its major features from the work of heterodox ascetics rather than from brāhmaṇic intellectuals and that the significant growth of Indian medicine took place in early Buddhist monastic establishments. In addition to forming the basis for a new history of medicine in ancient India, these findings should importantly advance our understanding both of the transmission and authorization of certain forms of knowledge through the *śāstra*s and of the social history of Buddhism in India and throughout Asia. This presentation of Indian medical history will likely stimulate controversy, particularly among those who ascribe the origins of *āyurveda* to traditional brāhmaṇic orthodoxy, and thereby contribute to a better and deeper understanding of India's medical heritage.

It is now my pleasant duty to thank those who have helped to bring this work to fruition. NIH Grant LM 04514 from the National Library of Medicine provided substantial funding for research and writing of the book from 1986 to 1988, and a Research Assistance Grant from the American Academy of Religion facilitated completion of the project. I deeply appreciate the support afforded me from these two institutions.

I am also grateful to several individuals who read all or part of the manuscript and made valuable comments. The foremost authority on *āyurveda*, Jan Meulenbeld, formerly of the State University of Groningen, made critical comments that corrected several errors and contributed to tighter arguments. J. W. de Jong of the Australian National University and Richard Gombrich of Oxford University, drawing on their deep knowledge of Indology and Buddhism, offered several insightful suggestions that improved the work's scholarly content. Stanley Insler of Yale University and David Pingree of Brown University caused me to rethink several arguments pertinent to my thesis. James Waltz of Eastern Michigan University read the entire text, raising scholarly questions and offering suggestions that greatly enhanced the book's general presentation. The views and ideas of these individuals, although not always adopted, were always considered and contributed to a surer grasp of the issues involved and a better understanding of the texts and their contents used throughout the book. To these and all individuals who looked at parts of the manuscript or who listened to me talk about the project and offered suggestions along the way, I am most grateful. Finally, I appreciate the constant support and encouragement given me by my wife, Adriana Berger, during the latter stages of the project.

London K.G.Z.
September 1989

Contents

Introduction

This study investigates the development of Indian medicine in the crucial but neglected period from about 800 to 100 B.C.E. Prior to that period, the magico-religious healing tradition of the early Vedic period (ca. 1700–800 B.C.E.) flourished; subsequent to it, the empirico-rational tradition of *āyurveda* found expression in the classical treatises of Caraka, Bhela, and Suśruta (ca. 200 B.C.E–400 C.E.). Critical examination of pieces of information from a wide variety of sources and use of historical and socioreligious approaches produce a more comprehensive and more plausible picture of ancient Indian medical history than that provided by philologically or philosophically based investigations.

Several facets of this deeper and fuller knowledge of Indian medical history may be mentioned here. Previously only two distinct phases of ancient Indian medical history could be delineated: the magico-religious healing of the early Vedic period, exemplified in the medical charms of the *Atharvaveda* and certain healing hymns of the *Ṛgveda*, and the empirico-rational medicine of *āyurveda*, expounded in the classical treatises of Caraka, Bhela, and Suśruta, which probably began to take shape in the centuries immediately preceding and after the turn of the common era. Early Vedic medicine was characterized by demon-caused diseases and magical rituals involving the recitation of potent charms and the application of efficacious amulets to exorcise disease demons and ward off their further attacks. Āyurvedic medicine, on the contrary, encompassed a sophisticated scholastic medical system recorded in specialized medical textbooks that present a distinctive medical epistemology relying essentially on empiricism followed by explanations of observable phenomena. The sharp contrast between these two medical traditions inescapably focuses attention on the intervening period in an attempt to elucidate the historical development of ancient Indian medicine.

Unlike the Indian traditions of physical science, which remained indissolubly connected with brāhmaṇic orthodoxy because of their essential role in the Vedic ritual, Indian medicine was never used by the Hindu sacrificial cults and was not a product of the orthodox brāhmaṇic intellectual tradition. However, medical knowledge was subjected to an assimilation process common among dominant orthodox religious-intellectual systems, whereby new information undergoes sufficient modification and adaptation to permit its integration into an established corpus of specialized knowledge.[1] Thus traditional brāhmaṇic sources recount the origin of Indian medicine through a lineage of divine, semidivine, and venerable transmitters. In these works, the sacred knowledge of healing began with the Hindu god Brahmā, who told it to Prajāpati, the Lord of Beings, who passed it on to the Aśvins, the physicians of the gods, who revealed it to Indra, the commander in chief of the gods. Indra then taught the esoteric medical knowledge to the divine Dhanvantari, who appeared in the form of Divodāsa, King of Kāśī (Banāras), from whom Suśruta learned it and transmitted it to mankind in his *Suśruta Saṃhitā*. Indra also revealed it to the sage Bharadvāja, who communicated it to other sages, including Ātreya Punarvasu, who trained six disciples among whom were Agniveśa and Bhela. For the benefit of humankind, Agniveśa composed a medical treatise reworked by his student Caraka and subsequently redacted by Dṛḍhabala, resulting in the *Caraka Saṃhitā*. Bhela transmitted his master's words in the *Bhela Saṃhitā*, which has come down to us in fragmented and corrupted form. This legendary beginning and passage of medicine, Jean Filliozat explains, is paradigmatic of Hindu science.[2]

This study contends that the traditional account of Indian medicine is merely the result of a later Hinduization process applied to a fundamentally heterodox body of knowledge in order to render it orthodox. Variations in an established corpus of specialized knowledge occur when different intellectual traditions become involved in the transmission and codification of specialized information, resulting in new conceptual models that often preserve remnants of prior paradigms, which serve to authorize the new forms of knowledge by establishing connections with the past. This apparently is what happened in Indian medicine. Heterodox ascetic intellectuals accumulated, systematized, and transmitted a body of medical lore that was later assimilated and processed by Brāhmaṇs to fit into an orthodox Weltanschauung.

Stated another way, during the approximately eight centuries of this "heterodox" period of Indian medical history, there was a radical shift in the way healers conceived of disease and its cure. The philosopher of science Thomas Kuhn attributes such intellectual transformations in Western science to "paradigm shifts," whereby revolutionary new modes of conceptualizing and explaining scientific problems wholly replaced old, established patterns of thought, resulting in scientific advancements.[3] Cautiously applied, Kuhn's theory of paradigm shifts can also be useful in understanding the revolutionary nature of the transition in Indian medicine from a magico-religious to an empirico-rational medical paradigm. Medical science in India deviates from Kuhn's rule by maintaining aspects of previous medical practices in the radically new approach to healing. Traditional techniques of magical medicine were never completely abandoned, but were assimilated into the new system of *āyurveda.* A revolutionary new medical paradigm replaced the old one and accommodated aspects of the former in the effort to render the entire system orthodox by demonstrating continuity. The example of Indian medicine in every other respect closely approximates Kuhn's model. The involvement of the heterodox ascetics provided the epistemological basis for a radically new way of conceiving mankind's afflictions and their cures, resulting in a shift to a new medical paradigm that assimilated forms of the old model as it became an orthodox science.

Another emphasis of this study is therefore to elucidate more thoroughly the contributions of heterodox ascetic renunciants, particularly Buddhists, to the developments of Indian medicine in the period of transition. A close scrutiny of the sources from the ninth century B.C.E. to the beginning of the common era reveals that medical practitioners were denigrated by the brāhmaṇic hierarchy and excluded from orthodox ritual cults because of their pollution from contact with impure peoples. Finding acceptance among the communities of heterodox ascetic renunciants and mendicants who did not censure their philosophies, practices, and associations, these healers, like the knowledge-seeking ascetics, wandered the countryside performing cures and acquiring new medicines, treatments, and medical information, and eventually became indistinguishable from the ascetics with whom they were in close contact. A vast storehouse of medical knowledge soon developed among these wandering physicians, who, unhindered by brāhmaṇic strictures and taboos, began to conceive an empirically and rationally based medical epistemology with which to codify and systematize this body of efficacious medical information. Fitting into the Buddha's key teaching of the Middle Way between the extremes of world indulgence and self-denial, healing became part of Buddhism by providing the means to maintain a healthy bodily state characterized by an equilibrium both within the organism and between the body and its environment. Portions of the repository of medical lore were codified in the early monastic rules, thereby giving rise to a Buddhist monastic medical tradition.[4] The symbiotic relationship between Buddhism and medicine facilitated the spread of Buddhism in India, led to the teaching of medicine in the large Indian conglomerate monasteries, and assisted the acceptance of Buddhism in other parts of Asia. Probably during the early centuries of the common era, Hinduism assimilated the storehouse of medical knowledge into its socioreligious intellectual tradition and by the application of an orthodox veneer rendered it a brāhmaṇic science.

Medicine in the Buddhist monastery receives special attention because, like the Christian monasteries and nunneries of the European Middle Ages, communities of Buddhist monks and nuns played a significant role in the institutionalization of medicine. Indeed, an understanding of the social history of Buddhism is incomplete without a full elucidation of Buddhism's involvement in the healing arts. The codification of medical practices within the monastic rules accomplished perhaps the first systematization of Indian medical knowledge and probably provided the model for later handbooks of medical practice; the monk-healers' extension of medical care to the populace and the appearance of specialized monastic structures serving as hospices and infirmaries increased the popularity of Buddhism and ensured ongoing support of the monasteries by the laity; and the integration of medicine into the curricula of major monastic universities made it a scholastic discipline. In India and elsewhere in Asia, Buddhism throughout its history maintained a close relationship with the healing arts, held healers in high esteem, and perhaps best exemplified the efficacious blending of medicine and religion. Even today, monks in the Buddhist countries of South and Southeast Asia treat patients for a variety of illnesses, and monasteries often include infirmaries in their compounds. This long-lasting union of religion and medicine in Buddhism contrasts sharply with their separation in Western civilization.

Moreover, this study seeks to advance the methodology employed in studying Indian medicine and to stimulate additional investigations, resulting in further contributions. Scholarship on Indian medical history from the early Vedic to the āyurvedic period is meager, and what exists is inadequate. From the narrow perspective of a historian of Indian philosophy, Debiprasad Chattopadhyaya's *Science and Society in Ancient India*[5] explores the evolution of classical *āyurveda*. The author rightly argues that Indian medical epistemology is fundamentally opposed to brāhmaṇic ideology and that the classical medical treatises of Caraka and Suśruta result from a Hindu grafting process whereby orthodox brāhmaṇic ideals were superimposed onto a medical framework. Unfortunately, his study is defective because it offers little historical evidence or explanation for the origin of the medical epistemology and pays slight attention to the wealth of medical information in Buddhist sources. Jyotir Mitra's *Critical Appraisal of Āyurvedic Material in Buddhist Literature*[6] provides a helpful codification of medical references in the Buddhist canon, but its lack of either theoretical framework or critical analysis renders it unsuitable for understanding the historical development of Indian medicine. In contrast, the pages that follow trace the development of India's antique medical tradition from its beginning in the early Vedic period to the formation of the classical medical treatises, focusing on the period of shifting paradigms and the role played by the ascetic traditions and Buddhism in facilitating the transition. Furthermore, a historical-philological methodology is used to elucidate the evolution of ancient Indian medicine and the characteristics of the Buddhist monastic medical tradition.

The structure of the study is simply grasped. Part I examines the history of Indian medicine from its beginnings in the early Vedic period to its absorption into the brāhmaṇic intellectual system as an orthodox science and traces the role of medicine in Indo-Tibetan and Chinese Buddhism. Drawing on a wide range of literary, archaeological, and secondary sources, this portion of the study provides a comprehensive picture of ancient Indian medicine in its socioreligious context. Part II investigates the constitution of Buddhist monastic medicine as presented in stories of sick monks whose treatments the Buddha purportedly sanctioned on a case-by-case basis. The medical content of each story is philologically analyzed and compared with the classical medical treatises to identify more precisely the relationships between Buddhist monastic and āyurvedic medicine and to obtain a fuller understanding of the common repository of heterodox ascetic medical knowledge that both exploited. Principal sources for Part II are the Pāli Vinaya Piṭaka, with the commentary of the Śrī Laṅkan Buddhist savant Buddhaghosa (fifth century C.E.), and the earliest extant Sanskrit medical treatises: *Caraka Saṃhitā*, with the commentary of Cakrapāṇidatta (eleventh century C.E.); *Suśruta Saṃhitā*, with the commentaries of Ḍalhaṇa (twelfth century C.E.) and Gayadāsa (eleventh century C.E.) on the *Nidānasthāna*; and *Bhela Saṃhitā*, which has no extant commentary. Concluding appendices supply a case-by-case study, using the philological analysis followed in Part II, of the cures performed by the lay physician Jīvaka Komārabhacca (Appendix I) and a glossary of Pāli and Sanskrit plant names with Linnaean and modern equivalents (Appendix II).

Several other potential lines of inquiry have not been pursued in this work. A detailed investigation of Jaina monastic sources is not undertaken because medicine generally played an insignificant role in Jaina monasticism. Likewise, no attempt is made to discuss cross-cultural comparisons between either Indian and Hellenistic or Indian and Chinese medical doctrines and etiological theories because the data available at present do not support conclusions, although the peripatetic life-style of the early roaming physicians suggests interesting possibilities.[7] Furthermore, given the association between the ascetic and medical traditions, Yoga's precise relationship to the healing arts deserves proper investigation. More immediately productive might be further examination of Prākrit literature against the background of Indian medical history provided in this study, which could conceivably supply additional useful information pertaining to Indian medical heritage.

I

THE EVOLUTION OF CLASSICAL INDIAN MEDICINE

1

The Beginnings of Indian Medicine: Magico-Religious Healing

An accurate picture of the growth and development of medicine in India must begin with an examination of the earliest phase of Indian medicine using available information derived from the archaeological remains of the Harappan culture and from the literary sources of the early Vedic period. Although the former only suggest a form of healing that may be characterized as fundamentally magical, the latter paint a clear picture of a predominance of magico-religious medicine.

Speculations on Harappan Medicine

In the absence of deciphered literary remains, artifacts from the chalcolithic sites of the Harappan culture, or Indus valley civilization (fl. 2300–1700 B.C.E.), located in present-day Pakistān and western India, provide the scholar's principal data.[1] Most cities of this culture were situated on the banks of the Indus River and its tributaries. Among them, two large centers, named after the modern villages near them, stand out as major urban establishments: Mohenjo-dāro in the south and Harappā in the north. The culture apparently was highly developed, stable, and not preoccupied with warfare. Its cities appear to have been well planned and its society stratified, with artisans and farmers supporting the upper levels. Seemingly, animism was the dominant form of religious belief, with both wild and tame animals receiving reverence. Numerous bull images, perhaps symbolizing the fertilizing heavens, suggest that that animal was a particularly sacred object of worship. Additionally, the earth appears to have been worshiped in mother goddess form. Religious practices of the Indus valley people may have included purification rites, magic, and fire rituals.[2]

Water as the source of all life and the most powerful purifying agent apparently held a significant place in the minds and lives of the Harappan people. The so-called Great Bath in the citadel area of Mohenjo-dāro might have served as a special place for the higher orders to bathe and perform religious ablutions in sacred waters. Such ritual purity might have been closely linked to the notion of public hygiene, for the design of the cities and houses suggests that sanitation was a principal concern. Moreover, Indus cities included a bath and toilet in almost every house, a drainage system to remove wastes from homes, and covered sewers centered in the streets to convey them away from residential areas.

Cultic worship of an Earth Mother goddess, depicted in numerous terra-cotta figurines, was probably practiced among agrarian segments of the society. She was likely the focus of regular domestic rites, during which prayers were recited before her image and substances were burned in small holders on both sides of the image's head. By consistently demonstrating reverence to the earth in the form of a mother goddess and to the sky in the form of a bull, devotees undoubtedly hoped to ensure the production of abundant cereal-grain crops. Vegetation, represented in the form of a female deity, could also have been an object of cultic worship, receiving reverence connected to the cult of the Earth Mother goddess. The paucity of images depicting worship of a female plant divinity, however, suggests that rituals devoted to her took place at prescribed times and probably in designated sacred locations for specific purposes. Seals and small rectangular sealings discovered at various Harappan sites show what appears to be reverence for plants and symbolically depict the birth of plants from the Earth Mother goddess, and their harvest. Large supplies of grains, stored in large granaries situated in citadels that perhaps served as the cities' religious and administrative centers, could sustain the population during periods of drought, thereby contributing to the stability of the society. Moreover, vegetation was probably a fundamental medicinal substance whose healing efficacy was guaranteed by worship of the plant goddess. Later evidence shows that vegetal products have always held a central place in Indian materia medica.

Asceticism also seems to have been practiced by certain Harappan individuals. Several seals and sealings unearthed at Harappā and Mohenjo-dāro portray a figure seated in a traditional yogic *āsana* (posture), perhaps *utkaṭikāsana*, the sitting posture in which the soles and heels of the feet are brought together and the legs form right angles.

Representations of the Indus ascetics strongly suggest shamans or medicine men. Individuals are depicted wearing elaborate ritual costumes, including bangles on the arms and a horned headdress. They are seated above the ground on a small platform or table, either alone or surrounded by animals, positioned in a type of yogic *āsana* with what appears at least on one seal to be an erect penis.[3] If the Indus ascetics were shamans, then they performed ritualistic magical healings, the principal function of shamans throughout the world. The shaman heals by means of magical rituals, including such elements as ecstatic dance, magical flight, the use of potent herbs and amulets, the recitation of incantations, and exorcisms.[4]

Like the medicine of the great civilizations of Egypt and Mesopotamia, the healing system of the Harappan culture was inextricably connected with the culture's religious beliefs and practices and probably based predominantly on magic. Using as a guide comparisons with the medicine of Mesopotamia, with which the Harappans had trade contact, and of ancient Egypt,[5] together with the impressions gathered from Indus remains, we may speculate that certain aspects of Harappan medical practices likely involved séances conducted by medicine men of a shamanistic type. In their healing rituals, such practitioners probably used, among other things, plants (worshiped to ensure their healing efficacy), recitation of powerful incantations (*mantra*s), and the performance of ritualistic dances and other activities in order to exorcise diseases believed to result from demonic possession. Apparently unique to the Indus valley civilization, however, was an emphasis on personal hygiene, supported by religious practices of ritual purification involving bathing and ablutions and by well-developed urban systems of waste disposal. Several skeletal remains suggest that the religio-surgical operation of trepanation could have been performed in Harappan-dominated areas, but no conclusive evidence for it is yet known.[6]

Magico-religious Medicine of the Early Vedic Period

The Harappan civilization slowly declined as changes in the courses and level of the rivers coupled with erratic climatic patterns reduced water supplies and crop production and led eventually to economic collapse and social decay. In this weakened condition, it could have quickly succumbed at around 1500 B.C.E. to charioteering Āryan invaders originating perhaps in Central Asia. Contact between the urban Harappans and the seminomadic Āryans occasioned a mixture of ideas not yet fully understood. Thus although Āryan literary sources supply a fairly accurate portrait of medicine in the early Vedic period, detailed analysis of the connections and continuity between Āryan and Harappan medical systems awaits further evidence from the investigations of archaeologists and the efforts of linguists to decipher the Indus script.

At the time of their invasions, the Āryans possessed in oral Sanskrit form at least the beginnings of a sacred scripture, known as the Veda, which contains numerous references to medicine. The oldest portion of the Vedic scriptures, the *Ṛgveda*, includes passages that look back to a time prior to the Āryan incursion into the Indian subcontinent, but was compiled in its final form not much before 800 B.C.E.[7] The *Ṛgveda* was the liturgical book of the *hótṛ*s, Āryan priests whose principal function was originally to perform sacrifices to the gods; hence the work is essentially a collection of hymns devoted to various divinities, mostly deified natural phenomena, who form the Vedic pantheon. Several of the hymns center on healing deities, most importantly the Aśvins, twin horsemen known as the "physicians of the gods." The sacred scripture also contains verses that make passing references to diseases, usually of demonic origin, and to other deities who sometimes engaged in healing activities. A unique hymn in the tenth book, whose language and subject matter suggest that it is later than the majority of hymns in the corpus, is devoted exclusively to the efficacy of healing plants.[8]

A slightly later Sanskrit text is the *Atharvaveda*,[9] the book of the *átharvan*s, fire priests or magicians skilled in the performance of magical rites. Much of the material in this treatise is at least as old as the *Ṛgveda* and combines priestly religious notions with more secular concerns perhaps reflective of mainstream, popular culture. It is essentially a book of magical charms, spells, imprecations, and incantations for numerous ends, including protection against demons and sorcerers, restoring the affection of a mistress whose love has grown cold, securing the birth of children, expiating sins, and succeeding in battle, trade, and even gambling. Because the *Atharvaveda* incorporates a significant number of charms devoted to the removal of disease, it is the principal source for medicine during the early Vedic period. However, the specifically medical hymns are not collected in any single part of the work but are scattered throughout its twenty books, primarily in books 1 through 9 and in book 19.

Around the third century B.C.E., the *atharvan* tradition produced a ritual treatise, the *Kauśika Sūtra*, which explains the various rites during which the hymns of the *Atharvaveda* are to be recited.[10] Of limited value for determining the actions of the early Atharvavedic rites, its passages often reflect ritual procedures artificially constructed to fit the contents of the older text and, in the medical section, contain ideas representative of the later medicine of classical *āyurveda* rather than the medicine of the early Vedic age.

The following description of medicine during this period derives from my earlier detailed study of the subject.[11] Textual evidence indicates that Vedic medicine, like that of contemporary societies, was fundamentally a system of healing based on magic. Disease was believed to be produced by demonic or malevolent forces when they attacked and entered the bodies of their victims, causing the manifestation of morbid bodily conditions. These assaults were occasioned by the breach of certain taboos, by imprecations against the gods, or by witchcraft and sorcery. Broken bones and wounds were understood to result from accidents or warfare, but demonically motivated noxious insects also sometimes inflicted injury on humans. These views concerning the origin of human infirmities imply a twofold understanding of disease: while maladies affecting the inside of the body were caused by one or several invisible disease demons, injuries, wounds, and similar afflictions on the outside of the body resulted from observable misfortunes, attacks in battle, or visible pests. A fundamental association existed between the ailment and its perceived cause: internal, invisible maladies came from unseen causes; external, visible afflictions derived from seen causes. This is an example of the most elementary kind of sympathetic or associative magic.

Internal diseases were of two basic types: those that included and manifested symptoms of *yákṣma* (consumption) and *takmán* (fever); and those that exhibited neither of the two symptoms and included *ámīvā*, tetanus (?), ascites, insanity, worms, urine retention, and perhaps constipation. External afflictions involved broken bones, flesh wounds, loss of blood, loss of hair, and various types of skin disorders. A third category of morbid bodily conditions resulted from different types of poison. Perceived as being both inside the body and on its surface, poison had both visible and invisible causes: animals, battle, and witchcraft. The long and renowned tradition of Indian toxicology might well have its origins in the medicine of the early Vedic period.[12]

Indian diagnosis did not include divination, as did that practiced in the medical traditions of ancient Egypt and Mesopotamia. Determining the cause of one's affliction was accomplished, rather, by isolating and identifying dominant and recurring symptoms, many of which were considered to be separate demonic entities. This technique, unique to Vedic medicine, exhibits a strong emphasis on observation and may mark the beginnings of empiricism and the Indian penchant for classification.

According to Vedic medicine, a healthy or sound body was one from which any one of the disease-causing agents was absent. Thus there were numerous sound states, depending on the particular demon or affliction eliminated, rather than one general healthy condition of mind and body.[13]

Elimination of morbid bodily conditions was the domain of the healer (*bhiṣáj*), a male professional whose special craft was the mending of the sick by the removal of disease demons and the repair of an injured part of the body. He, like a shaman, a medicine man, or an ecstatic, possessed knowledge of medicinal and potent plants, could recite the appropriate incantations, and could enter into trance states, during which he trembled and danced.

Evidence from both the *Ṛgveda* and the *Atharvaveda* indicates that the Vedic healer possessed only a superficial understanding of human anatomy. Although healers were learned and knew the Vedic hymns appropriate to their special craft, they were not members of the priestly orders and so could not participate in the Vedic sacrificial functions, which were exclusively part of the priestly domain. Their knowledge of anatomy could have been acquired either directly by observation of sacrifices at the large Vedic celebrations or indirectly through the routine execution of their practice. The precise techniques employed in the ritual immolation of animals and perhaps also of humans gave rise to an extensive Vedic vocabulary of the corporeal members. However, the Vedic scriptures record anatomical parts solely for the purpose of preserving the exact procedures to be followed during a sacred ritual, not for the purpose of obtaining a thorough understanding of the body and its functions. These terms, preserved in the ritual and exegetical treatises of the *Brāhmaṇa*s (ca. 900–500 B.C.E.), have survived in the later āyurvedic medical literature, with essentially their same meanings. The evidence shows that in the Vedic period the purpose of learning the bodily parts was religious rather than scientific.[14] Nevertheless, the anatomical information offered in these religious documents approaches scientific precision, resulting in quite detailed descriptions of the corporeal parts.

The medical hymns of the *Atharvaveda* indicate that therapy was carried out by means of specialized healing rites, in the course of which *mantra*s or charms from the sacred text were recited and various activities characteristic of a healing séance were performed. References to apotropaic materials commonly occur in the charms. Plants and their products constituted the predominant types of amulets. They were consecrated and made especially potent by the recitation of the appropriate *mantra*(s). Other talismans included the deciduous horns of animals. These magically potent substances, together with the powerful charms of the *Atharvaveda*, formed the healer's arsenal with which he engaged in ritual battle to expel demons and to protect victims from further attacks. There are also indications that fragrant plants were burned to ward off malevolent forces and to make the site of the healing rite auspicious. These rites took place at prescribed times of the day, month, and year, implying that knowledge of the auspicious times was also an integral part of Vedic medicine. Therapeutics for several external afflictions and poisoning reflect certain empirical and rational approaches to healing. They often involved primitive forms of surgery and even hydrotherapy, nearly always executed in conjunction with a magical healing rite.

The recounting of mythological events pertaining mostly to diseases and plants was central to many of the healing charms. Moreover, recitations of these myths were essential components of the healing ritual itself. Perhaps more than any other aspect of medicine in the early Vedic period, the retelling of these stories in a ritual setting links the ancient medical system to the Āryan Vedic religion. Mythology, Mircea Eliade posits, has in religious ritual the function of transporting the sacred proceeding back to the primordial time, the *illud tempus.* For the believer, it provides continuity and ensures a successful outcome.[15]

The Atharvavedic medical charms present a mythology specific to the function of healing and thus unique in Vedic literature. Divinities of the traditional Vedic pantheon are mentioned, usually as subordinates to the dominant Atharvavedic deity, often because their characteristics resemble those of the god being addressed (e.g., Rudra as lightning and thunder occurs in the context of Atharvavedic *takmán* [fever associated with the onset of the monsoon rains]). Most of the deities named in the charms are either malevolent demons of disease or benevolent plants and their products. The existence of a medical mythology points to a particular Vedic tradition that had the principal function of restoring members of the society to physical and mental wholesomeness and of maintaining them in this condition through specialized rituals. Its practitioners were not part of the priestly sacrificial tradition but freely borrowed elements from it to accomplish their ends. Evidence from the medical mythology of the *Atharvaveda* suggests a conscious effort by savants of this tradition to combine aspects of the priestly and the medical traditions, perhaps to authorize the latter in a society dominated by the former and to make the healers equivalent to the sacrificial priests at least within the arena of medical ritual.

The Homologization of Botanical Knowledge

The general trend of merging elements of the dominant priestly sacrificial tradition into medical rituals is perhaps best exemplified in the mythologies of healing plants, which illuminate the intellectual process of incorporating one tradition into another and reveal an apparent homologization of botanical knowledge. The two myths occur in the *Atharvaveda* and involve the healing plant goddess Arundhatī, used in the treatment of wounds and fractures, and the healing plant god Kuṣṭha, the remedy par excellence for fever (*takmán*). Arundhatī as *lākṣā́*, the Sanskrit term for "lac," probably refers to this resinous substance produced on trees by the so-called lac insect.[16] Traditional *āyurveda* recognized its medical efficacy among its many applications.[17] The plant Kuṣṭha is probably the same as the aromatic costus, native to Kaśmīr, and known to have been exported from India in the ancient spice trade. Its medical applications have been well known to traditional Indian physicians.[18]

The various epithets and forms by which Arundhatī was known are fundamental to her mythology. They are unique and probably derive from a tradition different from that represented in the *Ṛgveda*. As *arundhatī́*, she was a perennial (*jīvalā́*), harmless (*naghāriṣā́*), life-giving (*jīvantī*) plant with a honey-sweet flower, and the queen of all plants; as *silācī́*, she was said to arise from mount, and creep along trees, and was called the gods' sister, whose mother is the night, father is the cloud, and grandfather is death (Yama), from the blood of whose horse she was born; as *lākṣā́*, she had hairlike bristles on her stems and was the sister of the waters, whose self has become the wind; as *viṣāṇakā́*, she was said to have arisen from the fathers' root; and as *pippalī́*, she was buried by the Asuras and dug up again by the gods.[19] Although precise identification of the plant *arundhatī́* eludes us for the moment, certain key terms revealing this plant goddess's characteristics indicate her significant role in the wider scheme of Vedic medical botany.

The mythology of Kuṣṭha, although particular to the *Atharvaveda*, contains elements special to the *Ṛgveda* and, when compared with the mythology of Arundhatī, points to the homologization of the sacrificial and medical botanical traditions. *Kúṣṭha* was considered to be a divine, aromatic plant with all-pervading strength, the medicine for all diseases and the choicest among the herbs. He was thrice born from various divinities and was known by ancient, venerable men. He was acquired by the gods from the divine place where there is the appearance of immortality, the third heaven from earth where there is the seat of the gods, the *aśvatthá* tree, and where golden boats with golden rigging sail. He was called Soma's brother and, like Soma, grew high in the Himavant Mountains, the birthplace of eagles. The golden boats transported *kúṣṭha* to his earthly mountain home, whence he was brought to the people in the East and bartered. These parts of the myth of *kúṣṭha* link him closely to the Ṛgvedic *sóma*, the divine plant employed in the priestly sacrificial cults. The myth of the god Soma recounts that his place was in the highest (third) heaven and that he grew high in the mountains, where he was brought from his heavenly abode by an eagle. Later he is identified with the moon, which in the myth of Kuṣṭha is understood to be the golden boat.[20]

The name of Kuṣṭha's father was *jīvantá* (life-giving); of his mother, *jīvalā́* (perennial). His epithets include *naghamārá* (nondestroying) and *naghā́riṣa* (harmless). The goddess Arundhatī also, as noted above, was called "perennial," "life-giving," and "harmless." This nomenclature, together with Kuṣṭha's association with Soma, suggests a conscious effort to homologize a Ṛgvedic botanical tradition dominated by a male plant divinity with a medical-botanical tradition of plant goddesses particular to the *Atharvaveda.*[21] Although a detailed examination of the relevant Ṛgvedic and Atharvavedic passages is required, a perusal of each suggests that in the older strata of the *Ṛgveda* terms for various plants occurring in the masculine gender are often connected with Soma, while in the *Atharvaveda* plant names are more numerous and, when found in the masculine, are often associated with Soma or Kuṣṭha and, when in the feminine, with Arundhatī or one of her epithets.

Persistent occurrence of myths involving female healing plant divinities in the *Atharvaveda* implies a special reverence given to plants in the form of goddesses. The variety of vegetation and the descriptions of vegetal parts and plant habitats expressed in the medical hymns indicate an astute knowledge of the local flora and a rudimentary understanding of plant taxonomy.[22] Absence of references to such an elaborate medical botany in the hymns of the *Ṛgveda*[23] points to the presence of a tradition of professionals specializing in magic and medicine, particular to the *Atharvaveda.* Unlike the priests of the sacrificial cults, these ritualists were experts in the manipulation of the spirits, acquired extensive knowledge of the local flora necessary for their particular healing craft, and integrated elements of the sacrificial tradition into their sacred knowledge preserved in the *Atharvaveda.* Characteristics of this magico-religious medical tradition imply contact with an indigenous tradition already possessing knowledge of the surrounding flora and its usefulness to mankind. The Harappan cult of the Earth Mother goddess, the worship of plants in the form of female divinities, the representations of ascetic figures indicative of shamans or medicine men are perhaps examples of an indigenous healing tradition resembling the preceding portrait of early Vedic medicine. Although a precise connection between Harappan and Vedic medicine cannot at present be made, the similarities between them are highly suggestive.

The magical medicine of the Veda never completely disappeared in

India. It survived in classical āyurvedic medicine principally in the treatment of ailments that have Vedic parallels, in the cures for childhood diseases, and in remedies involving the elimination of malevolent entities.[24] Magical forms of medicine also occur in early Buddhist medicine[25] and characterize Mahāyāna Buddhist medicine.[26] Aspects of magico-religious medicine were practiced alongside the techniques of the more empirico-rational tradition of *āyurveda*, which revered Vedic medicine as the original form of brāhmaṇic medicine. The incorporation of medicine reflective of the early Vedic period into the classical āyurvedic treatises helped to authorize the newer medicine by establishing its links to the older healing tradition of the sacred Hindu Veda. The following chapters trace the evolution of Indian medicine as its paradigm shifted from a magico-religious system to one dominated by an empirical and rational epistemology, and focus on the crucial role ascetic traditions played in facilitating this transition.

2

Heterodox Asceticism and the Rise of Empirico-rational Medicine

We now embark on an investigation of later Vedic scriptures and materials deriving from nonbrāhmaṇic literary sources to trace the shift from magico-religious to empirico-rational medicine in ancient India. Crucial to this transition were the social status of the physician and the role of heterodox ascetic traditions.

Social Status and Ritual (Im)purity of Healers

Healers of the early Vedic period provided a necessary service to the ancient Āryan culture of India. The literary tradition of the *Atharvaveda*, which preserves knowledge of their specialized craft, indicates that healing and its practitioners were outside the general purview of the sacrificial cults and their dominant priestly ritualists but were comparable to the sacrificial priests in their particular sphere of ritual healing and respected for the special skills and knowledge they possessed. The precise social role of these masters of healing, however, is not easily ascribed. In a hymn of the *Ṛgveda*, they appear in the middle of a threefold list of skilled professionals that included carpenters (*tákṣan*), healers (*bhiṣáj*), and priests (*brahmán*).[1] Like the uneducated carpenters, healers repaired what was injured or broken, and like the learned priests, they commanded esoteric knowledge. Moreover, their skill in specialized healing rituals and knowledge of potent healing charms and incantations made them comparable to the ritualists (*vípra*) and priests (*brāhmaṇá*) of the sacrificial cults.[2] Physicians were a particular group of professionals who combined the craftsmanship of the woodwright with the intellection of the priest.

They were respected but were never granted a seat among the ritualists of the sacrificial cults. Because their mythological counterparts, the Aśvin twins, physicians to the gods, were praised in the Ṛgvedic hymns for the healing feats they performed, Debiprasad Chattopadhyaya wrongly concluded that physicians in the early Vedic period were highly esteemed,[3] an error resulting from his sole reliance on mythological references in the *Ṛgveda* rather than a more comprehensive picture derived from both the *Ṛgveda* and the *Atharvaveda.*

Subsequent literary works, particularly the late *Saṃhitā*s and early *Brāhmaṇa*s from the later Vedic period (ca. 900–500 B.C.E.), indicate that physicians and medicine were denigrated by the priestly hierarchy, who rebuked the physicians for their impurity and their associations with all sorts of people. A passage from the *Taittirīya Saṃhitā* provides striking evidence of the priestly contempt for physicians, here exemplified by the mythical Aśvins:

> The head of the sacrifice was cutoff; the gods spoke to the Aśvins: "you two are indeed physicians, [therefore] replace this head of the sacrifice." The two replied: "let us choose a choice [thing]; let now a ladleful [of Soma] be drawn here also for us." They drew this Aśvin portion [of Soma] for those two; thereupon, verily, the two replaced the head of the sacrifice; [hence] when the Aśvin portion is drawn, [it is] for the restoration of the sacrifice. The gods spoke to those two: "these two physicians, who roam with humans, [are] very impure." Therefore, medicine is not to be practiced by a Brāhmaṇ, for he, who is a physician [*bhiṣaj*], [is] impure, unfit for the sacrifice. Having purified those two with the Bahiṣpavamāna [Stotra],[4] [the gods], drew this Aśvin portion for them. Therefore, when the Bahiṣpavamāna has been chanted, the Aśvin portion is drawn. On account of that, the Bahiṣpavamāna should be reverently performed by the one who knows thus; verily the means of purification is the Bahiṣpavamāna; indeed, he purifies himself. [The gods] deposited the healing [powers] of those two in three places: a third in fire [Agni], a third in the waters, [and] a third in the Brāhmaṇ caste. Therefore, having placed the water vessel to one side [and] having sat down to the right of a Brāhmaṇ, one should practice medicine. To be sure, as much medicine as he practices by this means, his work is accomplished.[5]

Careful separation and examination of the mythical and human strands in this passage offer insight into the basis for brāhmaṇic attitudes toward healers and their art. From the mythological perspective, the Aśvins were at one time prohibited from partaking of the sacred Soma draught because they were in contact with impure mortal beings. The gods, however, required the services of the divine physicians to cure the sacrificial victim by replacing its head. In exchange for this service, the twin physicians requested and received the special favor of becoming coequal consumers of Soma. This mythical event produced two consequences in the human realm: the establishment of a rite of purification for physicians, involving the chanting of the Bahiṣpavamāna Stotra, in the Soma sacrifice; and the placement of divine healing powers in fire, in water, and in the Brāhmaṇ caste.

From the perspective of the brāhmaṇic social structure, it is similarly evident that physicians were considered to be impure and polluting, that they were excluded from the sacrificial rites, and that Brāhmaṇs were prohibited from practicing the art of healing. However, if individuals underwent a purification rite involving the Bahiṣpavamāna, they could then be accepted at the sacrifice and practice medicine in the brāhmaṇic setting. Moreover, strict ritualistic procedures, involving the philosophical, religious, and medically potent elements of fire, water, and the Brāhmaṇ priest (the last for his knowledge of healing *mantras*), had to be followed to render the physicians' medical procedures effective. A variant of the myth states that the physician should place a water vessel to one side and sit down to the right of the Brāhmaṇ, but omits mention of the element of fire. Other variants, however, mention all three elements, stating that the Brāhmaṇ should be given reverence at the edge of the sacrificial fire and that the waters should be made healing through the recitation of an incantation.[6]

Regarding the statement that physicians in general are polluting and therefore excluded from the brāhmaṇic sacrificial and social system, the *Śatapatha Brāhmaṇa* also confirms that physicians (i.e., the Aśvins) were impure because of their roaming among and constant contact with humans in the course of performing cures.[7] This attitude persisted in India and is found in the later law books that repeat passages from the Laws of Manu, stating that physicians (*cikitsaka, bhiṣaj*) must be avoided at sacrifices and that the food given by physicians is, as it were, pus (*pūya*) and blood (*śoṇita*) and is not to be consumed.[8]

Several explanations have been suggested for priestly defaming of physicians. Vedic scholar Maurice Bloomfield superciliously attributes it to the intellectual's proper evaluation of the "wretched hocus-pocus" of physicians and their craft[9]—a view that cannot now be taken seriously. Jean Filliozat claims that it expressed rivalry between the two Vedic schools of the Taittirīyas and the Carakas (implicitly connected with medicine through the name of the *Caraka Saṃhitā*).[10] Chattopadhyaya has shown this view to be untenable and asserts that the root cause for priestly contempt of physicians derived from a clash of philosophical perspectives between medicine's fundamental empiricism and the priestly ideology that emphasized esoteric knowledge. The Brāhmaṇs' philosophical outlook, he posits, led to "the stricture on direct knowledge of nature," which is fully expressed in the Upaniṣads and which the physicians systematically disregarded.[11]

A socioreligious perspective may provide additional insights. The paucity of references to medicine and healers in the literature of the late Vedic period inhibits conclusiveness but reveals that by that time defamation of the healing arts was already in place. Even in the early Vedic period, physicians were outside the pale of the Āryan sacrificial cults probably because of their association with the *Atharvaveda*, not yet considered a principal *śruti* (revealed) scripture. Moreover, their frequent travels beyond the frontiers of Āryan society in order to acquire the rich pharmacopoeia mentioned in the *Atharvaveda* brought them into frequent contact with non-Āryan peoples. Although physicians obtained from these outsiders much new and valuable knowledge pertaining to their special craft, these encounters caused them to be widely perceived as inferior beings polluted by contact with impure people. This attitude evidently existed from the early Vedic period but received articulation only in the later *Brāhmaṇa*s, which provided the orthodox brāhmaṇic means for accepting healers and consecrating their services. Their contact with non-Āryans might well have given rise to an empirical orientation that became, as Chattopadhaya correctly points out, antagonistic to brāhmaṇic orthodoxy in the later Vedic period.

Shunning of physicians and excluding them from the brāhmaṇic social structure and religious activities imply that they existed outside the manistream of society, probably organized into sects, and roamed the countryside, as indicated by the phrase "roving physicians" (*cāraṇavaidya*), the title of a lost recension of the *Atharvaveda.*[12] They earned their livelihood by administering cures and increased their knowledge by keen observation and by exchanging medical data with other healers whom they encountered along the way, for the āyurvedic medical tradition strongly encouraged discussions and debates with other physicians.[13] They were naturally indifferent, if not antagonistic, to brāhmaṇic orthodoxy because of their exclusion, and their special knowledge was in this late Vedic period not yet accepted as part of the orthodox brāhmaṇic intellectual tradition.

Several centuries after the beginning of the common era, however, when medical treatises were becoming established in their present forms, the physicians and their healing art became part of the brāhmaṇically based Hindu religious tradition. This is clearly evident from the introductory portions of these works, which recount how medicine was transmitted to humans from the Hindu god Brahmā.[14] The surgical treatise of *Suśruta Saṃhitā* refers to perhaps an Epic or a Purāṇic version of the myth in which the Vedic twin physicians, Aśvins, replaced the head of the sacrificial victim and offers a radical reinterpretation of the myth as an explanation for counting the branch of major surgery (*śalya*) first in the enumeration of the eightfold system of medicine:

> Indeed, this branch [i.e., *śalya*] is first because of its previous [use] in curing wounds from attacks and because of its [use] in the joining of the head of the sacrifice, for, as it is said: "the had of the sacrifice was cutoff by Rudra. Thereupon, the gods approached the Aśvin twins [saying,] 'lords, you two will become the best among us; [therefore,] the head of the sacrifice is to be joined by you two lords.' Those two replied, 'let it be so.' Then, for the sake of those two, the gods appeased Indra with a share of the sacrifice. The head of the sacrifice was joined by those two."[15] And also among the eight divisions of *āyurveda*, this [*śalya* division] is indeed supposed superior because it does its work quickly, because of its use of blunt and sharp surgical instruments, and cauterization with heat and caustic medicines, and because of its congruence with all other divisions. Therefore, this [*śalya* division] is eternal, sacred, heavenly, famous, providing longevity, and also giving a livelihood. Brahmā proclaimed [it first], Prajāpati [learned it] from him, the Aśvin twins from him, Indra from the two Aśvins, and I [i.e., Dhanvantari] [learned it] from Indra. Here now it is to be taught by me [i.e., Dhanvantari] to those desiring [it] for the good of humankind.[16]

Similarly, the medical authors attribute the origin of their science to the textual tradition of the *Atharvaveda*, which, by the time of the final redaction of the medical treatises during the early centuries of the common era, was included with the *Ṛgveda*, *Sāmaveda*, and *Yajurveda* in a fourfold classification of sacred Hindu texts. This form of authorization is exemplified in a passage from the *Caraka Saṃhitā*:

> Therefore, by the physician who has inquired about [which Veda an āyurvedic practitioner should follow, (previous verse)], devotion to the *Atharvaveda* is ordered from among the four [Vedas]: *Ṛgveda*, *Sāmaveda*, *Yajurveda*, and *Atharvaveda*. For it is stated that the sacred knowledge of the fire priests [*atharvans*] is medical science because [it] encompasses giving gifts, invoking blessings, sacrifice to deities, offering oblations, auspicious observances, giving burnt offerings, restraint of the mind, and recitation of magico-religious utterances, and so on; and medical science is taught for the benefit of long life.[17]

Mention of the *Atharvaveda* here, contrasted with the other Vedic

*Saṃhitā*s, is significant, for this textual tradition preserved the earliest medical lore. Generally considered to contain material at least as old as the *Ṛgveda*, the *Atharvaveda* was nevertheless held by Brāhmaṇs to be of a lower order than the other three Vedic texts, a view maintained until quite late. The first inclusion of the *Atharvaveda* in a Hindu tradition of sacred *śruti* (revealed) texts occurs probably after the beginning of the common era. In the medical literature, however, the *Atharvaveda* is given full authority as an orthodox treatise alongside the other sacred texts of the priestly order, and its inclusion serves to authorize the medical tradition in the Hindu cultural and religious milieu.

These developments signify a radical shift in the evolution of the medical tradition within the religious traditions of ancient India away from the previous disdainful attitudes of the Brāhmaṇs toward the medical practitioners and their art. This reflects what Chattopadhyaya has shown to be a late Hindu layer in the medical *śāstra*-literature, superimposed on an already well-established body of medical knowledge.[18] The occurrence of this phenomenon may indeed correspond to the fourth or fifth century of the present era, when Buddhism was declining in India and the brāhmaṇic religious tradition was making its resurgence through a radical reorientation of Brāhmaṇism. Although considered to be extremely polluting and defiling, medicine was now included among the Hindu sciences and came under brāhmaṇic religious influences, perhaps out of necessity as the need for the healing and care of the sick and injured cut across the existing social and religious barriers or, more likely, as a result of the general process of brāhmaṇic assimilation.

During the centuries intervening between Vedic medicine and the absorption of Indian medicine into brāhmaṇic orthodoxy (ca. eighth century B.C.E. to early centuries C.E.), the medical paradigm dramatically shifted from a magico-religious to an empirico-rational approach to healing. This transition occurred largely because of close associations between medicine and the heterodox ascetic traditions of ancient India. The shunned medical specialists—wandering the countryside, administering cures to all who required (and could pay for) them, and closely studying the world around them while exchanging valuable information with their fellow healers—understandably gravitated toward those sharing a similar alienation and outlook: the orthodox mendicants and the heterodox wandering ascetics who had abandoned society to seek liberation from the endless cycle of birth, death, and rebirth, and who were quite indifferent or even antagonistic to the brāhmaṇic orthodoxy of class and ritualism based on sacrifice to gods of the Vedic pantheon.[19] The heterodox ascetics, generally known as *śramaṇa*s (Pāli *samaṇa*), seem also to have had a penchant for more empirical and rational modes of thought. As A. K. Warder suggests, they attempted to find explanations of the universe and of life by their own efforts and reasoning powers and were particularly interested in the natural sciences.[20] The physicians began to associate freely with the *śramaṇa*s, among whom medicine developed and flourished, and found Buddhist *śramaṇa*s most favorably disposed toward the healing arts. The exact date of the śramaṇic sects is uncertain, but they were common from the sixth century B.C.E. because Buddhists, Jainas, and Ājīvikas were all known to be *śramaṇa*s.[21]

Ascetic Physicians and Their Medical Art

Literature of the heterodox religious movements and accounts of foreign travelers contain numerous references to healers and medicine and reveal the heterodox ascetics' familiarity with and involvement in Indian medical lore. The Jaina canonical text *Vyākhyāprajñapti* (or *Bhagavatī Sūtra*) recounts a visit to Makkhali Gosāla, the founder of the Ājīvikas, by six wandering ascetic philosophers (*disācara*s).[22] One of them was a certain Aggivesāyaṇa, who, according to A. L. Basham, bears a clan name perhaps connected to Agniveśa, the semilegendary physician on whose teachings the *Caraka Saṃhitā* is based.[23] Associations between the art of healing and the *śramaṇa*s also occur in the Buddhist Pāli canon. The *Brahmajālasuttanta* of the *Dīghanikāya* enumerates the livelihoods (*jīvikā*) in which certain *samaṇa*s and Brāhmaṇs engaged for their subsistence. The Buddhists considered these to be brutish arts, avoided by the *samaṇa* Gotama (Buddha). One section lists the following medical practices: causing virility and impotence; giving emetics, purges, and purges of the upper and lower parts of the body and of the head; administering oil in the ears, refreshing the eyes; nasal therapy; applying collyria and ointments; ophthalmology; major surgery; pediatrics; giving of root medicines; and administration and evacuation of herbal remedies.[24] As several of these medical therapies are part of early Buddhist medicine outlined in the section on medicine of the *Mahāvagga*,[25] the condemnation here was against accepting payment for performing any of the services (Buddhists were forbidden remuneration for their actions). These therapies are representative of medical procedures generally prescribed in the early āyurvedic treatises.

Significant evidence for the connection between the medical arts and the wandering ascetics (*śramaṇa*s) derives from the accounts of the Greek historian Megasthenes (fl. 300 B.C.E.), a special ambassador sent by the first Seleucus to the court of Candragupta Maurya at Pāṭaliputra (modern Patna). His observations are communicated by the historian and geographer Strabo (ca. 64 B.C.E.–21 C.E.):

> And with regard to the Garmanes [i.e., *Śramaṇa*s], [Megasthenes] says that some, the most esteemed, are called Hylobii [i.e., forest dwellers], who live in the forests, [existing] on leaves and wild fruits, [wearing] clothing of tree bark, without [indulging in] sexual intercourse and wine. [He says that] they associate with the kings who, through messengers, inquire about the causes [of things]; and who, through those [Hylobii], serve and petition the divinity. And after the Hylobii [Megasthenes says that] the physicians come second in [so far as] honor; and [that they are] philosophers, as it were, concerning mankind, frugal, but not living off the land, who sustain themselves with rice and barley groats, all of which, [he says,] the one who is begged and who welcomes them in hospitality, supplies to them; and [he says that] they are able to bring about multiple offspring, male offspring and female offspring, through the art of preparing and using drugs; but they accomplish healing through grains for the most part, not through drugs; and of the drugs [he says that] the most highly esteemed are the ointments and the plasters; but the rest have much that is harmful. And [he says that] both the latter and the former [*śramaṇa*s] practice endurance, both actively and inactively, so that they can continue being fixed in one posture the whole day; and there are others who are prophetic, skilled in the use of incantations, and skilled in the words and customs associated with the "departed," and who go begging through both villages and cities; on the other hand [there are] others who are more attractive than these and more urbane; but even they themselves do not refrain from the common rumors about Hades, insofar as it seems to [tend] toward piety and holiness. And [he says that] women as well as [men] study philosophy with some of them, and the women also abstain from sex.[26]

The passage clearly points to a connection between the physicians (*ἰατρικοί*) and the *śramaṇa*s (Γαρμᾶνας), recognizing the former as a subgroup of the latter. Moreover, the Greek account characterizes the rational therapeutics of both the early Buddhist monk-healers and early āyurvedic physicians. Megasthenes's depiction of the ascetic physicians as not living in the fields, being frugal, feeding themselves with rice and barley groats, begged from hospitable supporters, and practicing ascetic austerities fits numerous mendicant orders in ancient India, including Buddhist monks and nuns living the monastic life in the early *saṅgha*s, or monastic establishments.

The details of medical practice are even more revealing. The śramaṇic healers are said to effect their cures mostly through grain foods (*σιτία*), and when they employ drugs (*φάρμακα*), the most esteemed are ointments (ἐπίχριστα) and poultices (καταπλάσματα). Inherent in this distinction is the internal dietary use of foods and the external application of drugs, both of which are fundamental to the rational therapy (*yuktivyapāśraya*) of āyurvedic medicine.[27] The former helps to sustain and regulate the internal functions of the human organism by restoring a balance to the bodily elements, while the latter eradicates afflictions located on the body's surface.[28] Medical passages contained both in the Buddhist monastic code (Vinaya) and in the early āyurvedic treatises are replete with illustrations of the medicinal use of foods and the therapeutic application of remedies such as ointments and poultices.[29]

The observation that these *śramaṇa*-physicians were philosophers concerning mankind (περὶ τὸν ἄνθρωπον φιλοσόφους) discloses the philosophical orientation of medicine that required a materialist or naturalist perspective. Human beings, according to the ancient āyurvedic physicians, were the epitome of nature.[30] Proper understanding of nature (i.e., *svabhāvavāda*) required above all profound knowledge of the human species,[31] and Indian medical theoreticians placed paramount emphasis on direct observation as the proper means to know everything about mankind.[32] Empirical data they acquired were, in Chattopadhyaya's words, "then processed by rational analysis leading to a medical epistemology"[33] that is unique to Indian medicine. In contrast to the Greek tradition in which empiricism gave rise to natural history, the empiricism of the āyurvedic tradition, according to Francis Zimmermann, led to "the inclusion of taxonomy in pharmacy and the subordination of pharmacy to a complex interplay of savors and curative properties, which envelops the inventory of the flora and fauna in a welter of synonyms, redundancies, enumerations, divisions, and cross-references."[34]

Complete knowledge of humans and their relationship to their environment included an understanding of the causes of mankind's ailments. Indian medicine's inherent philosophical orientation led to theories about causes for mankind's afflictions. Although its exact origin cannot be determined, the etiology particular to Indian medicine is the three-humor (*tridoṣa*) theory. Nearly all the maladies plaguing humans are explained by means of three "peccant" humors, or *doṣa*s—wind, bile, and phlegm—either singly or in combination. The *doṣa*s are really specific waste products of digested food, occurring in quantities greater or lesser than needed to maintain normal health. They act as vitiators by disrupting the normal balance of the bodily elements (*dhātu*s), which in turn are modifications of the five basic elements (earth, air, fire, water, and ether) found in all of nature, and the resulting disequilibrium of the bodily elements produces disease. Their empirical orientation also led the medical theoreticians to include environmental factors, daily regimen, and external factors in their overall consideration of the causes of diseases.

Reference to this etiology occurs in the Pāli canon. While questioning the Buddha on the cause of mankind's suffering, the wandering ascetic Sīvaka, who may have been a physician, said that some *samaṇa*s and *brāhmaṇa*s claimed that suffering was caused only by previous acts (i.e., *kamma*). The Buddha said that that view was incorrect and explained that the cause of mankind's suffering was eightfold: "bile [*pitta*], phlegm [*semha*], wind [*vāta*] and [their] combination [*sannipāta*], changes of the seasons [*utu*], stress of unusual activities [*visamaparihāra*] [e.g., sitting or standing too long (i.e., excesses), going out hastily at night, or being bitten by a snake], external agency [*opakkamika*] [e.g., being arrested as a robber or an adulterer], and the result of (previous) actions [*kammavipāka*] is the eighth [*aṭṭami*]."[35]

The Buddha's eightfold formula of disease causation includes the three "peccant" humors and their combination, which is central to āyurvedic etiology. All four also figure significantly in the later Mahāyāna discussions of disease causation.[36] Elsewhere in the Pāli canon, a physician (*tikicchaka*) is known as one who administers purges and emetics for checking illnesses that arise from bile, phlegm, and wind.[37] The last four causes mentioned by the Buddha are external rather than internal factors helping to produce suffering, particularly morbid bodily conditions. Explanations of disease arising from the seasons (*ṛtu*), from unusual or irregular activities, objects, or foods (*viṣama*), and from past actions (*karman*) occur in the early medical treatises of Caraka and Suśruta.[38] External agency (*opakkamika*), which is Sanskrit *upakrama* and Pāli *upakkama* (*opakkama*, *opakkamika*), has the primary meaning "attacking suddenly" (root *kram* plus *upa*). It therefore could be equivalent to the *āgantu*, or external, category of disease causation in Indian medicine. According to the āyurvedic medical tradition, *āgantu* causes are generally violent and traumatic and involve injury to the body.[39]

The inclusion of past actions (*karman*, *kamma*) as a category of medical etiology is clearly quite old and deserves special attention. The notion that past actions contribute to an individual's overall physical state is, as Chattopadhyaya rightly argues, in conflict with the general empirico-rational physiology of Indian medicine. He erroneously infers from that observation, however, that any mention of *karman* as a cause of disease represents a later superimposition of religious theory onto an already established system of materialist medicine in an effort to render medicine religiously orthodox.[40] *Karman* is, however, counted as a factor in the eightfold enumeration of disease causation recognized by the Buddhists, which suggests that medical theoreticians, purely on the level of theory, accepted it quite early. The *Caraka Saṃhitā* mentions that a certain Bhadrakāpya was the principal proponent of the theory, but the context in which the passage occurs demonstrates that it was by no means universally followed.[41] Mitchell Weiss's study of the doctine of *karman* in medical literature indicates that in the *Suśruta Saṃhitā*, *karman* is very nearly ignored as an etiological category, while in the *Caraka Saṃhitā* its inclusion as a factor in the causation of disease is discussed purely at a theoretical rather than a practical level. *Karman* does nevertheless seem to have been taken seriously in sections of the *Caraka Saṃhitā* dealing with embryology.[42] A restatement of the canonical passage, occurring in the later Pāli Buddhist treatise *Milindapañha*, which dates from the early centuries of the common era,[43] confirms the Buddhists' long-held view that *karman*, or past action, was not the sole cause of mankind's suffering, asserting that that which arises as a result of *karman* is far less than that which arises from other causes.[44] Because the dominant explanations for the causes of disease stemmed from the empiricism and rationality of materialist physicians as confirmed by the Buddhists, *karman* is not, as Chattopadhyaya claims, an example of later brāhmaṇic grafting.

There are clear indications that an early form of materialist medical philosophy was widely attributed to *śramaṇa*-physicians whom Megasthenes characterized as "philosophers concerning mankind" and that medical presuppositions associated with rational medicine were already familiar to early Buddhists. Through their common philosophical orientation, the link between the ascetic physicians, the Buddhists, and the āyurvedic doctors is further established.

Megasthenes's observation that the *śramaṇa*-physicians produced many offspring, male and female, through the art of preparing and using drugs (*δύνασοαι···διὰ φαρμακευτικῆς*) does not find a parallel in Pāli Buddhist records but does reflect a teaching current in later Mahāyāna scripture. A Chinese fragment of the *Kāśyaparṣiproktastrīcikitsāsūtra* (*Sūtra of Gynecology Taught by the Ṛṣi Kāśyapa*), focusing on the therapeutics for women during various stages of their pregnancy, bears a very close similarity to the section "concerning the aphorisms of procreation" (*jātisūtrīya*) in the anatomical books of the *Caraka*, *Bhela*, and *Kāśyapa Saṃhitā*s. The techniques involve the prescription of specific drugs during each of the ten or twelve months of the fetus's gestation.[45] The *Suśruta Saṃhitā* prescribes similar practices, firmly established in ancient Indian medical lore, for assuring the birth of healthy offspring of either sex in its chapter in the book of anatomy concerning "the purity of the sperm and the egg" (*śukraśoṇitaśuddhi*).[46]

In another section of his *Geography*, Strabo, referring to certain unidentified historians of India, again connects the *śramaṇa*s with the healing arts:

> In classifying philosophers, [the writers on Indian] set the Pramnai [i.e., *śramaṇa*s] in opposition to the Brachmanes [i.e., Brāhmaṇs]. [The Pramnai] are captious and fond of cross-questioning; and [they say that] the Brachmanes practice natural philosophy and astronomy, but they are derided by the Pramnai as charlatans and fools. And [they say that] some are called mountain dwelling, others naked, and others urban and neighboring, and [the] the mountain-dwelling [Pramnai] use [i.e., wear] hides of deer and have leather pouches, full of roots and drugs, claiming to practice medicine with sorcery, spells, and amulets.[47]

The mountain-dwelling Pramnai (*πράμναι*) in this passage differ from the *śramaṇa*-physicians described by Megasthenes. The healing of the Pramnai *śamaṇa*s is magico-religious, using sorcery (*γοητεία*), spells (*ἐπῳδαί*), and amulets (*περιάπται*), and reminiscent of the early Vedic medical tradition reflected in the *Atharvaveda*. This form of healing is, on the whole, contrary to the empirical and rational medicine of the early Buddhist and āyurvedic literature, in which references to magical techniques are rare.[48] The *Caraka Saṃhitā* does nevertheless recognize it as one of the three forms of therapy. It corresponds to the category of therapy based on the recourse to divine entities (*daivavyapāśraya*), which involved, among other practices, the recitation of *mantra*s; the use of vegetal and stone amulets; auspicious observances; the giving of offerings, gifts, and burnt offerings; restraint of the mind; atonement; invoking blessings; sacrifice to deities; prostration to the gods; and pilgrimage.[49] This form of medicine is nearly identical to that which the *Caraka Saṃhitā* attributes to the fire priests of the *Atharvaveda*.[50]

The testimony of this Greek source suggests that healing practices relying principally on magic were also part of certain śramaṇic traditions. Magical medicine is characteristic of the medical lore of the early Vedic period, when practitioners relied on magico-religious rather than on empirico-rational principles. Quite possibly, aspects of magico-religious healing, originating with the *átharvan*s, were also preserved by the *śramaṇa* groups and, through them, were transmitted into the early Buddhist and āyurvedic medical traditions. Specific knowledge of individual religious techniques may have varied, but the underlying precepts of magical and religious healing were maintained and incorporated into the general corpora of the emerging medical literature.

These two Greek accounts of the *śramaṇa*-physicians reveal that the principles and practices of both the empirico-rational and the magico-religious systems of Indian medicine were preserved among the heterodox groups of wandering mendicants. These śramaṇic traditions served as a storehouse of medical knowledge that the wanderers accumulated through observations of different forms of healing during their treks across the land and provided the medical doctrines utilized by both the early Buddhist monastic community and the professional medical practitioners.

The ancient Indian physicians (*vaidya* [Pāli *vejja*], *bhiṣaj* [Pāli *bhisakka*], *cikitsaka* [Pāli *tikicchaka*]), shunned and denigrated by the dominant orthodox brāhmaṇic society, found refuge in the less orthodox communities of renunciants and mendicants who did not censure their philosophies, practices, and associations and, as wanderers engaged in healing, gradually became indistinguishable from the *śramaṇas*. Only much later, perhaps out of the need for medical care or by a wholesale brāhmaṇic takeover, were these healers and their arts incorporated into the mainstream Hindu religious tradition, a subsequent development reflected in the medical treatises themselves by the superimposition of brāhmaṇic orthodoxy on an already well-established system of medical lore.

In this regard, it is interesting to look more closely at Caraka, the principal redactor of Agniveśa's medical treatise bearing his name. Controversy exists over the exact date of Caraka, with many scholars subscribing to the view of Sylvain Lévi, who, on the authority of the fifth-century C.E. Chinese translation of the *Sūtrālaṃkāra*, asserted that Caraka was a physician of King Kaniṣka, thus placing his treatise in the first or second century of the common era.[51] Chattopadhyaya, however, persuasively argues that the *Caraka Saṃhitā* in its original form was not the work of one person but "the compilation of medical knowledge of ancient roving physicians."[52] Because the treatise itself refers to different approaches to medicine and mentions numerous medical traditions, Chattopadhyaya's claim rings true.[53] Moreover, the word *caraka* itself, a masculine noun from the root *car* (to wander), means a wanderer or an ascetic and aptly fits members of *śramaṇa* groups, the repositories of ancient medical lore. The treatise *Caraka Saṃhitā*, then, might likely refer to the "Compilation (of Medical Knowledge) of the Wanderers [i.e., *śramaṇas*]." Caraka might also have been the name of a certain *śramaṇa*-physician who was court physician to the Kuṣāṇa king, Kaniṣka, and may also have participated in editing and compiling an already existing body of medical lore, but the evidence thus far marshaled does not confirm this.

The Ascetic's Knowledge of the Human Body

The connection between heterodox, particularly Buddhist, asceticism and medicine is perhaps best illustrated through anatomy. The approach of the early Buddhists and the physicians to an understanding of the human body reflects both a commitment to materialism through empiricism and rationality and a firm rejection of brāhmaṇic orthodoxy.

Early Buddhist literature of the Pāli canon contains a fairly accurate picture of the gross anatomical parts of the human body. An enumeration of the corporeal members and the means by which they were ascertained are found in sections of the Sutta Piṭaka treating the ascetic discipline, rather than in the medical portions of the Vinaya Piṭaka. In the *Mahāsatipaṭṭhānasuttanta* of the *Dīghanikāya*, the four intents of contemplation (*cattāro satipaṭṭhana*) are detailed. The first of these was the human body (*kāya*) in all its parts, aspects, and impurities. The monk was to endeavor, through persistent contemplation, to realize the fundamental impermanence of his physical and mental constitution by meditating on the body:

> And in addition, O monks, a monk contemplates this very body, up from the soles of the feet [and] down from the crown of the head, bound by skin [and] full of manifold impurities. "There is in this body [the following]: hair of the head, hair of the body, nails, teeth, skin, flesh, sinews, bones, bone marrow, kidney, heart, liver, pleura, spleen, lungs, bowels, intestines, stomach, excrement, bile, phlegm, pus, blood, sweat, fat, tears, grease, saliva, mucus, serous fluid, and urine...." There is in this body "earth element, water element, fire element, and wind element."[54]

This enumeration indicates a knowledge not only of the external but also of the major internal parts and exudations of the human body. Such an understanding of anatomy implies firsthand observation of the body. An important variant adds to this list the brain (*matthaluṅga*) inside the head,[55] demonstrating that fairly detailed investigations of organs located not only beneath the ribs but also inside the cranium were undertaken.

The means by which this anatomical knowledge was acquired is also expressed in the text. The monk should reflect on the body and learn its parts in the same way as a skilled butcher of cattle (*dakkha goghātaka*) or his pupil (*goghātakantevāsin*) does when he, having slaughtered the cow and divided it into parts, sits at the crossroads.[56] This method suggests that anatomical knowledge was acquired originally by observing the parts of an animal as they were cut away from the body and systematically arranged. By analogy, then, the knowledge gained from this dismemberment procedure was applied to the human body.

Another way of obtaining information about the human body involved persistently focusing and concentrating on a decomposing corpse that had been thrown on a charnel ground (*sīvathika*). The monk was to reflect on a putrefying body, dead from one to three days, becoming bloated and decaying, being devoured by animals, until its bones became bleached white and eventually turned to powder.[57]

Intense observation of decomposing bodies combined with knowledge of the anatomy of animals gave early Buddhist ascetics excellent means for obtaining an understanding of the gross internal and external structures of the human body. The religious and philosophical presuppositions of those engaged in this ascetic exercise were fundamentally conducive to the acquisition of this special knowledge. Buddhist ascetics, being *śramaṇa*s and nonbrāhmaṇic in their outlook, were not compelled to conform to orthodox brāhmaṇic rules pertaining to purity and pollution, which prevented Brāhmaṇs or members of the higher orders from undertaking such endeavors. Direct observation of a decaying corpse considered polluting by Brāhmaṇs and the upper classes was the best and most valid way to gain knowledge of human anatomy for the specific purpose of demonstrating the Buddhist doctrine of impermanence to the ascetic monks, but in addition, it afforded an empirical understanding of the human body.

A recent study examining the evolution of anatomical knowledge in ancient India from the point of view of the brāhmaṇic textual traditions shows that the ancient Indians initially acquired an understanding of anatomy through the Vedic sacrificial rites in which horses, cows, and probably humans were immolated according to a precise ritual procedure requiring the name of the bodily part to be recited while an offering was tossed in the sacred fire.[58] Enumerations of anatomical parts resulting from this rite are preserved in the texts of the *Brāhmaṇa*s. The slaughter of cows, outlined above in the Buddhist account, probably reflects techniques similar to those employed in the Vedic sacrificial ritual.

By the time of the early medical treatises in the early centuries of the common era, another technique for obtaining knowledge of the human body, involving a type of dissection, was employed and is detailed in the *Suśruta Saṃhitā*:

> Therefore, after having cleansed the corpse, there is to be a complete visual ascertainment of the limbs by the bearer of the knife [i.e., the surgeon] who desires a definite knowledge [of the body].
>
> For, if one should learn what is visually perceived and what is taught in the textbooks, then both together greatly increase one's understanding [of the human body].

> Therefore, after having removed the feces from the entrails, one should let decay a body with all its limbs intact, which has not been severely infected with poison, which has not suffered a prolonged illness, which is not badly injured, and which is not 100 years old [i.e., not overaged]. [The corpse,] wrapped in any of the coverings of *muñja* grass, tree bark, *kuśa* grass, or *śaṇa* hemp, and so on, [should be] placed in a cage (or a net) [and] bound in a driving stream, in a concealed spot; then, after seven nights, the completely putrid body should be removed [and laid out]. Thereupon, one should very gradually scrape off the layers of skin, and so on, by means of any of the bunches of vetiver grass [*uśīra*], coarse animal hair, bamboo, or *balvaja* grass and should identify with the eye all the various major and minor parts [of the body], both external and internal [ones], which have been mentioned previously [in this chapter]....
>
> The one skilled both in [the direct study of] the body and in the textbooks should be [known as] one whose purpose is clear. He should carry out his duties, removing his doubts by direct observation and by what he has heard [from his teacher].[59]

Because this method of dissection required the physician or student to come in contact with extremely impure and defiling substances, it very likely did not originate in the brāhmaṇic social and religious setting in which the tradition claims the early medical treatises, as we now have them, developed.

The opportunity for observing decomposing corpses deposited in rivers is not offered in the Pāli canon.[60] However, later reports from travelers to India confirm that this technique for disposing of the dead was practiced even by Buddhists of ancient India. The Chinese Buddhist pilgrim Hsüan-tsang, who traveled throughout India in the early seventh century C.E., relates three methods of disposing of the dead and of performing the burial rites: by cremation, by desertion, and by water. In the last technique, he states, "the body is thrown into deep flowing water and abandoned."[61] Likewise, the Muslim scholar and traveler Albīrūnī (ca. eleventh century C.E.) explains: "People relate that Buddha had ordered the bodies of the dead to be thrown into flowing water. Therefore, his followers, the Shamanians, throw their dead into rivers."[62] This reference clearly connects the depositing of the dead in rivers with the Buddhists and *śramaṇa*s, and implies that the practice dates from the time of the Buddha. It is possible that the *śramaṇa*-physicians, whose special knowledge was the human body, developed the technique of dissection taught in the *Suśruta Saṃhitā* in order to allow them to observe the body and to acquire a deeper understanding of its structure and individual parts. Such information pertaining to the acquisition of anatomical knowledge by the early

Buddhists contributes much to our understanding of the evolution of Indian anatomy and to the connections between *śramaṇa*, Buddhist, and āyurvedic medicine.

An empirical approach to learning human anatomy by dissection, involving direct, firsthand observation of the body, was fundamental to the āyurvedic medical knowledge and was also common to the Buddhist ascetics' quest to understand the human body. Although the individual purposes of the two traditions differed, their epistemologies and methodologies were identical and provided similar results. Just as Buddhist asceticism owed its origin to the *śramaṇa* tradition, so most probably the principle and practice of direct observation of the human body derived, like other aspects of the medical arts, from the heterodox *śramaṇa*s. The technique of dissection described in the medical treatise of Suśruta was likely incorporated into the larger body of medical teachings before the text crystallized into its extant form and probably occurred before the addition of the Hindu veneer.

The heterodox ascetic movements in ancient India provided a social and philosophical orientation found among both the early Buddhists and the early medical theoreticians. The healers, like the ascetics, were seekers of knowledge and outcastes, shunned by the orthodox Hindus. They wandered about, performing their cures and acquiring new medicine, treatments, and medical knowledge, and eventually become indistinguishable from the other *śramaṇa*s with whom they were in close contact. The healers were not necessarily ascetics, but many ascetics—for instance, the Buddhist monk-healers—might well have been physicians. A vast storehouse of medical knowledge developed among these śramaṇic physicians, supplying the Indian medical tradition with the precepts and practices of what has come to be known as *āyurveda*. The first documented codification of this medical lore took place as wandering ascetics assumed a more stationary existence, cloistered in the early Buddhist monasteries.

3

Medicine and Buddhist Monasticism

Medical knowledge was probably common to most of the fraternities of ascetic wanderers from the sixth century B.C.E., but it was among the Buddhists in particular that it became an integral part of religious doctrines and monastic discipline. Jainas obviously knew medical theories and practices, but because of the severity of their ascetic discipline, the cultivation and practice of techniques to remove and ease suffering operated essentially as a hindrance to spiritual progress.[1] Hence Jainas did not codify medicine in their monastic tradition.

Some have suggested that the Buddha's key teaching of the Four Noble Truths was based on a medical paradigm, whereby suffering, its cause, its suppression, and the method for its elimination correspond in medicine to disease, its cause, health, and the remedy. Although a fourfold division occurs in the *Caraka Saṃhitā*, it does not reflect the dominant mode of systematized medical knowledge and, moreover, its formulation differs from that of the Four Noble Truths: "The best physician, one fit to treat a king, is he whose knowledge is fourfold: [he knows] the cause [*hetu*], symptom [*liṅga*], cure [*praśamana*], [and] non-recurrence [*apunarbhava*] of diseases."[2] The insignificance of the fourfold division in the medical tradition and its conceptual variation from the Four Noble Truths render any medical analogy in the Buddha's original teaching untenable. Knowledge of medical theory and practice among the śramaṇic Buddhists, however, is indisputable, and the Buddhist *saṅgha*, or monastic community, soon became the principal vehicle for the preservation, advancement, and transmission of Indian medical lore.

Most authorities agree that the *saṅgha* began to take shape in the lifetime of the Buddha (ca. early to mid-fifth century B.C.E.).[3] At first, the term seemingly included all who accepted and followed the teaching of the Buddha in the so-called Bhikkhusaṅgha of the Four Quarters, whose bond of unity was the Buddha's teaching (*dhamma*), which appealed to males and females from all walks of life and whose purpose was to spread the Buddha's word.[4] Initially the Bhikkhusaṅgha was peripatetic, but eventually Buddhist monks (*bhikkhus*) and nuns (*bhikkhunīs*) settled down into fixed residences as temporary rain retreats (*vassa)* evolved into permanent monastic establishments, providing the necessities of life.

As the *saṅgha* evolved, regulations developed governing the cenobitical life. These ordinances, preserved in the Vinaya Piṭaka of the Pāli canon, detail every aspect of the lives of monks and nuns in the *saṅgha.* The guiding principle for *saṅgha* life, based on the Buddha's teaching of the Middle Way, states Richard Gombrich, was "that one should treat oneself well enough not to be distracted from spiritual life by hunger, and moderately enough not to be distracted by over-indulgence."[5] Providing the means to restore and maintain a healthy physical balance, medicine therefore was ideally suited to this philosophy of the Middle Way. Parts of the Vinaya refer to medical matters, and the rules pertaining to remedies and treatment codify a medical knowledge quite similar to that recorded in the early medical treatises. Chapters 5 and 6 offer a detailed analysis of this Buddhist medical information; this chapter, however, traces the growth of medicine in the Buddhist *saṅgha.*

Monk-Healers and the Monastery

In the early *saṅgha*, membership was quite unrestricted, and wanderers joined and left at will. These comings and goings increased the quantity of new information exchanged among the various *śramaṇa*s who happened to sojourn during the monsoon rains with the *bhikkhu*s who themselves were still active mendicants. Debates among the temporary residents were common and likely included topics related to medicine. (As already noted, debates among medical practitioners are advocated in the *Caraka Saṃhitā.*)[6] As fixed *saṅgha* establishments with permanent cenobites became more common, the knowledge discussed and exchanged was gradually accumulated, filtered, and codified, eventually becoming Buddhist doctrine. The record of the Buddhists' acquisition and development of a teaching pertaining to medicine and healing and fitting into the Buddha's doctrine of the Middle Way can be traced in the Pāli canon.

The tradition preserved in the Vinaya specifies that a new monk of the *saṅgha* was provided with four resources (*nissaya*): meals of morsels of food (acquired by begging), robes of rags from dustheaps, lodging at the foot of trees, and putrid urine (of cattle) as medicine (*pūtimuttabhesajja*).[7] Newly ordained nuns received the same resources save the lodging at the foot of trees, which was deemed unsafe for females.[8] These four bare necessities of life probably derive from an early stage of *saṅgha* development, when monks and nuns actively practiced the ideal of a wandering, nonpermanent life-style, because they were never made compulsory by the Buddha, and likely reflect ascetic practices of certain mendicants deemed to be too extreme for the Buddha's middle course.

The inclusion of a form of medicine as one of the essential life resources points to knowledge of techniques of healing among the wandering ascetics. Animal urine (usually from cattle) is included in the medical section of the Vinaya as the allowable treatment for snakebites[9] and is mentioned frequently in the early āyurvedic treatises as a principal ingredient in numerous recipes and therapies.[10] Because neither human donations nor injury to living beings was required to obtain it, animal urine was easily accessible to wanderers, and its medicinal use was widely known to all mendicant ascetics, including the Buddhists.

As a more settled existence in monastic establishments displaced the wandering life, four "possessions" (*parikkhāra*), modeled on the four "resources," superseded the latter as more appropriate for a stable and permanent life-style. However, medicines requisite in sickness (*gilānapaccayabhesajja*) remained among a monk's necessities and constituted one of these four possessions along with a robe, a begging bowl, and a bed-and-seat.[11]

Medicines included all those things necessary for the care of the sick, and were to be used only to ward off pain and to maintain health, never to give pleasure.[12] The chapter on medicines (*Bhesajjakkhandhaka*) of the *Mahāvagga* specifies the requisite medicines. Monks were permitted five basic medicines: clarified butter (*sappi*), fresh butter (*navanīta*), oil (*tela*), honey (*madhu*), and molasses (*phāṇita*). With the evolution of the *saṅgha* and the development of the Vinaya rules, the medicines grew into an entire pharmacopoeia, including numerous foods and incorporating culinary traditions, derived perhaps from the laity.[13] Early monastic establishments provided storerooms for foodstuffs (*kappiyabhūmi*),[14] and distribution of foods was assigned to various monks: for example, the *khajjabhājaka* (one who distributes solid food), the *yāgubhājaka* (one who distributes gruel), and the *phalabhājaka* (one who distributes fruits)[15] (both gruel and fruits were often used as medicines). Very likely, the monk responsible for the distribution of small necessities (*appamattakavissajjaka*) was in charge of the drugs and medical supplies of the early monastery.[16] Together, the pharmacopoeia and the medical practices constituted the requisites necessary in sickness.

Food and medicines were obtained by donations from pious laity, for monks and nuns were forbidden to work to buy these items or to produce them. With the evolution of cenobitical life, the practice of begging for alms was replaced by the acceptance of offerings brought to the monastic communities. Foods and foodstuffs were collected and divided into substantial and nonsubstantial materials, and the latter were further classified into medicines. Other requirements for medical therapy, delineated according to their usefulness in specific treatments, were similarly received from lay supporters during times of need.[17]

The development of the materia medica utilized in the Buddhist *saṅgha* seems to be based on a systematic classification of foods. A similar classification of medicines based on foods also appears in the early āyurvedic treatises. Gradual development of a Buddhist medical tradition by inclusion of rules pertaining to drugs and treatments for specific ailments relied on a codification of medical knowledge also present in large part in the early āyurvedic medical treatises. The congruence of these approaches strongly suggests a common origin for the medicine of both the Buddhist monastery and the early āyurvedic tradition.

The monk's role as a healer seems initially to have been restricted to the care of his fellow *bhikkhu*s. A story in the *Mahāvagga* relates how this function became institutionalized in the *saṅgha.* A certain monk suffering from a bowel disturbance (*kucchivikārābādha*) lay fallen in his own urine and feces (*muttakarīsa*). Because he was not useful (*akāraka*) to the monks in this condition, no one nursed (*upaṭṭheti*) him. However, the Buddha, coming upon this monk, nursed him and afterward propounded the following rule pertaining to the nursing of sick monks: "You, O *bhikkhu*s, have neither a mother nor a father who could nurse you. If, O *bhikkhu*s, you do not nurse one another, who, then, will nurse you? Whoever, O *bhikkhu*s, would nurse me, he should nurse the sick."[18]

To aid the monks in caring for their brethren, the *Mahāvagga* identifies the qualities (*aṅga*) of difficult and easy patients and of competent and incompetent nurses (*gilānupaṭṭhāka*).[19] The qualities of a patient easy to nurse are as follows: he does what is beneficial; he knows moderation in what is beneficial; he takes his medicine; he makes clear the affliction, as it arises, to the nurse and wishes him well, saying "it is progressing" as it progresses, "it is regressing" as it regresses, and "it is stable" as it stabilizes; and he endures the arising of bodily sensations that are painful, acute, sharp, severe, disagreeable, unpleasant, and destructive.[20] The qualities of a patient difficult to nurse are the exact opposites of these.[21] The qualities of a competent nurse to the sick are as follows: he is competent to provide medicine; he knows what is beneficial and nonbeneficial; he offers what is beneficial and takes away what is nonbeneficial; he nurses the sick with a kindly thought, not out of greed; he is not unwilling to remove feces, urine, mucus, or vomit; and he is competent to gladden, rejoice, rouse, and delight the sick from time to time with a story about Buddhist doctrine (*dhamma*).[22] The qualities of an incompetent nurse to the sick, again, are the exact opposites of these.[23]

The act of nursing a monk who fell ill was considered to be such a great service (*bahūpakāra*) that when the monk passed away, his begging bowl and robes were given to the one who nursed him, whether *bhikkhu* or novice (*sāmaṇera*), rather than going to the *saṅgha*. If both a novice and a *bhikkhu* nursed the sick, the bowl and robes were to be shared equally by the two.[24] An institution of monk-healers, utilizing medical doctrines codified under the monastic rules, thus evolved along with the medical system for the purpose of providing medical care to the sick in the *saṅgha*. The inclusion of special incentives in the form of material acquisitions given to those who attended the sick points to the importance of medical care in the Buddhist monastery,[25] but also implies that monks felt some apprehension about coming in contact with sick and diseased persons, reflecting perhaps orthodox brāhmaṇic notions pertaining to purity and pollution among certain monks who joined the *saṅgha*.

Early āyurvedic treatises also delineate the qualities (*guṇa*) of a physician, of a medical attendant, and of a patient. The *Caraka Saṃhitā* lists four qualities for each: those of a physician (*vaidya*) are excellence in medical knowledge, extensive practical experience, dexterity, and cleanliness:[26] of a medical attendant (*paricara*), knowledge of medical attendance, dexterity, affection, and cleanliness;[27] and of a patient (*ātura*), memory, obedience, fearlessness, and giving information about the disease.[28] In the *Suśruta Saṃhitā*, the number of qualities of each is not fixed: the physician (*bhiṣaj*) has correctly learned the medical treatises, has practical experience, acts for himself, has a quick hand, is clean and brave, has the surgical implements and medicines made ready, is able to make prompt decisions, is intelligent, resolute, and learned, and holds truth (*satya*) and duty (*dharma*) as highest principles:[29] the patient (*vyādhita*) as longevity and courage, is curable, possesses the drugs (required for treatment), has self-control, is pious (*āstika*), and is attentive to the commands of the physician;[30] and the medical attendant (*paricara*) is affectionate, irreproachable, strong, and intent on the care of the sick, carries out the commands of the physician, and does not tire.[31]

The qualities of the physician, patient, and medical attendant presented in the early medical treatises are not identical to those found in the Pāli Vinaya but share many points, significantly their codification in both the Buddhist and the āyurvedic literary traditions. However, they also exhibit important differences. Caraka's standardization of four qualities appears to be closer to lists of the Buddhist tradition, while Suśruta's random number of characteristics and use of stock brāhmaṇic words like *satya* (truth), *dharma* (duty), and *āstika* (pious) point to a late compilation of the virtues and illustrate the application of the subsequent Hindu veneer. The medical texts distinguish between the physician and the medical attendant or nurse, who relied on the instructions of the physician, while the *Mahāvagga* has only the attendant or nurse, who functioned alone as a healer. The separation and specialization of medical professions presented in the āyurvedic treatises imply a later development. Codification of the qualities of medical personnel and of patients finds an early formulation in the institutions of the early Buddhist *saṅgha*, reflecting the Buddhists' increased involvement with medicine and their preservation and development of medical doctrines that find parallels in the āyurvedic tradition, albeit with obvious brāhmaṇic accretions.

Lay women also provided medical service to both monks and nuns, especially in three ways identified by I. B. Horner: calling on sick nuns, occasionally visiting monasteries in the hope of finding invalids, and providing medicaments to monks and nuns. The lay devotee Suppiyā was best known for her work with sick monks and nuns.[32] She was so famous as a nurse to the *saṅgha* that, according to legend, as an extreme act of piety, she gave, in the absence of meat, a piece of flesh from her thigh to be used in broth for a sick monk.[33] There is no evidence that Buddhist nuns performed special functions of nursing and taking care of the sick; however, being generally obliged to follow the same rules as monks, nuns presumably had the responsibility of attending to the ill among their numbers.

Among the lay physicians (*vejja*, Skt. *vaidya*), several seem to have rendered medical service to the *bhikkhu*s without charge. A physician named Ākāsagotta of Rājagaha allegedly lanced a rectal fistula, a procedure monks were forbidden to perform.[34] The most renowned lay physician was Jīvaka Komārabhacca, who is described as providing free medical care to the Buddha and other monks and donating his mango grove at Rājagaha for use as a monastic community, named *Jīvakārāma*. His free medical service to monks is said to have attracted large numbers of people to join the order, thus creating problems for the *saṅgha*. Jīvaka's fame as a healer was widely known, and legends about his life and medical feats can be found in almost all versions of Buddhist scriptures.[35]

The Monastic Infirmary

Medicine and healing were integral parts of Buddhist monasticism from its inception. Initially, medical activities focused on the care and treatment of monks by fellow cenobites or by pious lay devotees, but from around the mid-third century B.C.E., there is evidence that the monk-healer and the monastery extended medical care to the population at large.

The second rock edict of King Aśoka (ca. 269–232 B.C.E.) at Girnar proclaims that everywhere in the kingdom medical treatment is to be provided to both humans and animals; medicinal herbs, roots, and fruits are to be imported and planted wherever they are not found; and wells are to be dug and trees planted along the paths.[36] This edict in no way proves that hospitals existed in India in the third century B.C.E.,[37] but suggests that the monk-healers' role of extending medical aid to the laity coincided with the spread of Buddhism during Aśoka's reign. The expansion of Buddhism and its monasteries from northeastern India generally followed the existing trade routes, and support for the new institutions ultimately came from the laity. Monastic establishments were built close to the actual avenues of trade and received funding largely from wealthy merchants who found the monasteries convenient "ports of call" and resting places (i.e., hospices) while on their long and wearisome treks across vast stretches of land.[38] A sure way to obtain the necessary donations was to offer more than spiritual guidance. Medical care and facilities to refresh the traveler were therefore provided for those who would reciprocate with donations to the *saṅgha*. This strategy proved extremely successful, as the *saṅgha* and its monastic establishments grew and flourished until the twelfth century C.E., when Buddhism was crushed by foreign invaders. A sixth-century C.E. inscription from the Duḍḍavihāra in Gujarāt, stating that the use of medicines and remedies was for all those who are sick, not only for the monks, lends support to this claim.[39]

Evidence of Buddhist hospitals or monastic structures devoted to the care and treatment of the sick is meager. Monks and nuns were usually treated in their own cells. However, a reference in the Pāli canon to a "hall of the sick" (*gilānasālā*), located at the Hall of the Peaked Gable (*kūṭāgārasālā*) in the Great Forest (*mahāvana*) near Vesāli, points to a structure in the monastery compound set aside for the care of sick brethren.[40] An inscription from Nāgārjunikoṇḍa, dating from the third century C.E., suggests that a health house for the care of those suffering and recovering from fever was part of this famous Buddhist monastery. The inscription, written in Sanskrit, is fragmented, but the relevant portion is as follows: [*śo*]*bhane vihāramukkhye vigatajvarālaye* ("in the splendid chief monastic house, [in] the abode of feverless"). The fever mentioned here could be mental distress or, more likely, any febrile disease, including malaria, the disease most dreaded by the Indians from the time of the *Atharvaveda*.[41]

In the fifth century C.E., the Chinese Buddhist pilgrim Fa-hsien related that at the city of Pāṭaliputra heads of Vaiśya families established houses for dispensing charity and medicine, to which the poor, the destitute, the maimed, the crippled, and the diseased could resort and receive every kind of help and where physicians would examine their diseases. They obtained the food, medicines, and decoctions they required and were made to feel at ease. When they had recovered, they departed.[42] Such a structure might have been the *ārogyavihāra* (health house) of the Buddhist monastery at Pāṭaliputra, discovered during the excavations at Kumrahār. The building, dating from around 300 to 450 C.E., had four rooms of varying size, with walls of fired-backed bricks and a floor of brick concrete. Among the debris unearthed at the site was an inscribed sealing that depicts a tree (*bodhi*?), with a conch shell on either side. The inscription found in the lower half of the sealing reads: *śrī ārogyavihāre bhikṣusaṅghasya* ("in the auspicious health house of the monastic community"). Two potasherds from the same debris bear respectively the inscriptions *(ā)rogyavihāre* (in the health house) and *(dha)nvantareḥ* (of Dhanvantari). The latter might be the title of the physician attached to the *ārogyavihāra*, who practiced medicine according to the surgical tradition of Dhanvantari, the divine source of the *Suśruta Saṃhitā*.[43] At a Buddhist site in Nepāl, an inscription dated at 604 C.E. (*śaka* 526) from a stele at Lele near Kāthmandu mentions a donation of land by a king for an *ārogyaśālā* (health house).[44] Mortars and pestles excavated from the monastery at Sarnāth, near Banāras, point to the use of a monastic structure to nurse and to treat the sick in the eighth and ninth centuries C.E.[45]

By the tenth century, Hindu religious centers had integrated medicine into their religious life and had established places to care for the sick and destitute. In southeastern Bengal, a copperplate inscription dating from about 930 C.E. mentions a grant from King Śrīcandra to provide for a physician to be attached to each of two *maṭha*s, or brāhmaṇic religious and educational institutions. In southern India, the Malkapuram stone pillar inscription from Āndhra Pradesh, dated 1261 to 1262 C.E., speaks of a grant from Viśveśvara Ācārya to a college attached to a *maṭha*. The college included a *prasūtiśālā* (maternity house), an *ārogyaśālā*, and a *viprasattra*, a place, provided with a physician (*vaidya*), where all people, regardless of caste, could be fed.[46]

An important eleventh-century (1069 C.E.) Tamil inscription from the

Viṣṇu temple of Veṅkaṭeśa-Perumāl at Tirumukkūḍal, in Tamilnādu, provides detailed information of a hospital attached to the temple. The record mentions the upkeep of the hospital; the medicines it stocked; the number of beds furnished for inpatients; and the funds given for a staff of nurses, a physician, a surgeon, and a compounder (apothecary), and for servants such as a washerman, a potter, and others who attended to the needs of the patient.[47] Finally, a fifteenth-century (1493 C.E.) Tamil inscription from the Hindu Raṅganāth temple at Śrīraṅgam, in Tamilnādu, mentions support for the restoration of an *ārogyaśālā* and the installation of a Dhanvantari shrine. This *ārogyaśālā* might have been originally constructed in the eleventh or twelfth century.[48]

Following the lead of Buddhist monastic medical institutions, Hindu religious centers at a later period established places to provide medical care and services to the sick and the poor. When Buddhism was submerged in India after 1200, these Hindu institutions seem to have assumed the responsibility for medical services previously provided by the Buddhist monasteries. Further research into the infirmaries and hospitals of ancient India should elucidate the process by which other religions became involved with medicine and the care of the helpless and ill, as well as illuminate the general role of religious centers in India's medical history.

Medical Education and the Monastery

The special system of monastic ordinances in the early Buddhist *saṅgha* provided a codification of medical knowledge and led to the development of monk-healers who treated both their brethren and their lay devotees and to a certain extent furnished care to the sick in special monastic structures or houses established and supported by the laity. Buddhism's involvement with medicine also gave rise to the teaching of medicine in the Buddhist monastic establishments after the beginning of the common era.

Among the early centers of education in ancient India, Taxila is perhaps the most renowned. The famous lay Buddhist physician Jīvaka Komārabhacca was to have obtained his medical education at Taxila by studying for seven years as apprentice to a physician who, according to Sanskrit and Tibetan accounts, was the semilegendary Ātreya, whose teachings formed the basis of the *Caraka Saṃhitā*.[49] Moreover, the *Kāmajātaka* (no. 467) relates a story of a king of Banāras who was cured by Bodhisatta who mastered all branches of learning, including medicine, at Taxila.[50] At the beginning of the common era, Taxila was regarded as the principal center of medical studies as well as education in the arts and the sciences and in traditional brāhmaṇic learning.[51] During the reign of the Kuṣāṇa kings in the northwest (first to third centuries C.E.), Taxila was a flourishing seat of Buddhism. Archaeological excavations from that region indicate that Buddhist monastic establishments existed there from the early Kuṣāṇa period until the sack of Taxila by the Huns in the fifth century C.E.[52] Taxila was therefore an excellent venue for the establishment of a symbiotic relationship between medical education and Buddhist monasticism.

At Nāgārjunikoṇḍa in the south, medical education may have been part of the curriculum of the Buddhist monastery in the early centuries of the common era. Next to the room for those recovering from fever, there existed in the monastery what some authorities claim was a type of classroom.[53] Given its proximity to the infirmary, both practical and theoretical education in medicine could well have taken place there.

The full blossoming of the Buddhist monastic "universities" at the large, conglomerate monasteries (*mahāvihāra*) apparently began toward the end of the Gupta dynasty (mid-sixth century C.E.). Describing the monastery at Nālandā in the early seventh century C.E., the Chinese Buddhist pilgrim Hsüan-tsang (in India from 629 to 645 C.E.) states that students from all quarters came to study with learned men at Nālandā. Those wishing to study at the monastery must first have been well educated in both new and old subjects and prepared to engage in intellectual debate.[54] The courses studied at Nālandā were the works of Mahāyāna and of the eighteen Buddhist schools and other works, including the Veda, logic, grammer and philology, medicine, Atharvavedic magic, Sāṃkhya, and several miscellaneous texts.[55] The three beginning with logic (*hetuvidyā*) and ending with medicine (*cikitsāvidyā*) form part of the five sciences (*vidyā*) of the traditional curriculum. Hsüan-tsang mentions that a student began studying these five sciences, along with the Buddhist textbooks (*śāstra*s), from the age of seven. With respect to medicine, he says that it embraced exorcising charms, medicine, the use of medicinal stones, needles, and moxa.[56] Brāhmaṇs, he states, studied the four Vedas, which begin with "longevity" (*āyurveda*), relating to the nourishment of life and keeping the constitution in order, include "worship" (*Yajurveda*) and "making even" (*Sāmaveda*), and end with the "arts" (*Atharvaveda*), relating to the various skilled crafts, incantations, and medicine.[57] The traditional brāhmaṇic enumeration of the four Vedas has the *Ṛgveda* or sacred verses in place of *āyurveda*. The inclusion of both *āyurveda* and *Atharvaveda* in the enumeration of sacred brāhmaṇic literature already reflects the incorporation of spurious traditions into the orthodox brāhmaṇic system, a process that coincides with the resurgence of Brāhmaṇism during the Gupta era. Medicine and magic that included aspects of healing were therefore made available to the higher orders of the orthodox society as subjects to be learned. It is noteworthy that certain forms of Indian medical practice appear to have been identical to traditional Chinese therapeutics.

During the latter half of the seventh century C.E., I-tsing, another Chinese Buddhist pilgrim who visited many Buddhist monasteries in India, also mentioned the study of the five sciences and explained that the science of medicine was composed of eight sections, the classical eight limbs (*aṣṭāṅga*):

> The first treats of all kinds of sores; the second, acupuncture for any disease above the neck; the third, of disease of the body; the fourth, of demonic disease; the fifth, of Agada medicine [i.e., antidotes]; the sixth, of the diseases of children; the seventh, of means of lengthening one's life; the eighth, of methods of invigorating the legs and body.[58]

The mention of acupuncture here refers to *śalya*, the branch of medicine that uses pointed instruments in surgical procedures on the eye, ear, nose, and throat. I-tsing goes on to compare Indian materia medica, certain medical treatments, the rules for giving medicine, and the avoidance of harmful medicinal treatment with the current medical practices in China.[59] Medical knowledge by the middle of the seventh century C.E. was codified as a system that was preserved in the classical medical treatises and well established in the curriculum as one of the five sciences taught in Buddhist monastic "universities."

By the fourteenth century, the curriculum involving the five sciences was an integral part of education in the Buddhist monasteries of Tibet, having been transported from the *mahāvihāra*s of India before their annihilation. The Tibetan Buddhist historian Bu-ston states that one of the principal texts used for medical instruction was Vāgbhaṭa's *Aṣṭāṅgahṛdaya Saṃhitā* (*The Collection of the Essence of the Eight Limbs* [of *Āyurveda*]), which was translated into Tibetan from Sanskrit. "As for the works on the medicine," he states,

> they teach four (topics), disease, cause of disease, medicament as antidote to disease, and the method of curing thereby; or—body, child, demon, body, upper body, dart, fang, (and) senile lust. These are said to be the eight parts. Having thus reduced (the material) to eight (sections)—pregnant woman, young child, its demoniac disease, body, i.e. interior trunk, upper body, i.e. head, wound by dart, wound by fang, a varan [monitor lizard]—They are like the *Aṣṭāṅga*(*hṛdayasaṃhitā*) in teaching it.[60]

In the beginning, medicine was part of an ascetically based religious movement, a portion of which became known as Buddhism. Medicine evolved along with the *saṅgha* and Buddhist monastery in India, became codified as part of the Buddhist scriptures, gave rise to the monk-healers and provided the basis for subsequent development of Buddhist monastic hospices and infirmaries, and finally became part of the standard curriculum in the Buddhist monastic "universities." When Buddhism began to spread to other parts of Asia, medical institutions and practices of the monastery went along as integral parts of the religious system. The traditional system of āyurvedic medicine owes much of its early systematization, preservation, and subsequent propagation to the ascetic Buddhists and their monastic institution.

4

Indian Medicine in Buddhism Beyond India

Buddhist canonical literature provides evidence of the importance of medicine, the commingling of different forms of healing within Buddhism, and the spread of Indian medical ideas to Tibet and parts of Central, East, and Southeast Asia. Except in Śrī Laṅkā and western Southeast Asia, where Theravāda Buddhism, based largely on the doctrines of the Pāli cannon, retained its primacy, Mahāyāna Buddhism exercised dominant influence. Mahāyāna, which was carried along Central Asian trade routes from India to China by the first century C.E. and became the official religion of Tibet during the eighth century, upheld the ideal of the bodhisattva who, acting out of compassion, deferred passage into final *nirvāṇa* to help other attain higher spiritual states and eventual release from the saṃsāric world of endless rebirth and in its approach to healing focused on providing relief to all individuals.[1] Its monastic medical doctrines, however, as indicated by extant versions of the monastic codes, remained essentially those codified in the Vinaya of the Pāli cannon. Nonmonastic medical documents of later Buddhism, nevertheless, provide evidence of influences from sectarian interests and indigenous medicine. Preeminent among these are magico-religious forms of healing associated with esoteric Buddhism, steeped in the use of magical utterances, incantations, and potent healing herbs, and with rituals and practices surrounding the cult of the bodhisattvas of healing. Some early Mahāyāna texts contain protective charms, *dhāraṇīs*, and *mantras*, short mnemonic strings of words devoted to a particular purpose and employed in magical ritual only by persons initiated in their proper use. The magical forms of healing probably owe much to influences from esoteric Tantras, which were studied in the northern Indian Buddhist monasteries from the sixth century C.E. However, esoteric practices reached their full expression in the Mantrayāna or Vajrayāna teaching of Tibetan Buddhism. The Tantras described, among other things, the acquisition of magical powers to cure disease, to ward off demonic attacks, and to stave off death. These practices appealed to the popular mind already inclined to such beliefs by the indigenous religious practices and attitudes of the quasi-Buddhist Bonpos. By demonstrating magical powers, Buddhist monks could win followers who requested their services in dealing with extrahuman forces. Through the manipulation of spirits, they could alter the normal course of transmigrations and reverse the effects of *karman*. In this way, magical healing became an integral part of Tibetan Buddhism.[2] Additionally, the cult of the Buddhas and bodhisattvas of healing marks a new development in the transmission of medical traditions of Buddhism from Central Asia to China. Along with treatises expounding magico-religious healings, the principles and practices of empirico-rational medicine as explicated in Indian *āyurveda* were exported to Tibet in the form of translations of Sanskrit medical treatises incorporated into the Tibetan Buddhist canon, known as the *Kanjur* and *Tanjur*.

This chapter first examines the monastic medicine represented in the Vinaya recension of the various Buddhist schools, compares the Pāli, Sanskrit, Tibetan, and Chinese versions of the legend of the physician Jīvaka Komārabhacca, and finally surveys medical information contained in nonmonastic Buddhist literature from India, Central Asia, Tibet, and China.

The Common Vinaya Core of Buddhist Monastic Medicine

Central to healing traditions of the Buddhist monastery are the medical sections preserved in the chapters (*skandhaka*, Pāli *khandhaka*) of the *Mahāvagga* of the Vinaya Piṭaka. The Vinaya exists in the recensions of six Buddhist schools: Sarvāstivāda, Dharmaguptaka, Mahīśāsaka, Mahāsāṃghika, Mūlasarvāstivāda, and Theravāda. Those of the first four are preserved in Chinese; the fifth is in Chinese, Tibetan, and partly in Sanskrit; and the sixth is in Pāli. Erich Frauwallner has examined all the *skandhaka*s of the extant Vinayas in a study containing much information useful in ascertaining the spread of Buddhist monastic and Indian medicine to parts of Asia beyond India; he concluded that in general the content of the Vinayas, including the inserted legends, is similar in all the recensions.[3] Attributing the insignificant differences among them to oral transmission, he states, "We must therefore accept a common basic work, from which the Vinaya texts... are derived."[4] The oldest stratum of the Vinaya includes the *skandhaka*s or "chapters," which he thinks "must have been composed shortly before or after the second [Buddhist] council [of Vaiśālī]... [in] the first half of the fourth century B.C."[5] The chapter on medicines (Pāli *Bhesajjakkhandhaka*, Skt. *Bhaiṣajyavastu*) is part of the oldest and most original section of the Vinaya, and therefore represents the earliest form of Buddhist monastic medicine.[6]

Fundamental agreement exists among these versions in the enumeration of materia medica, containing the five basic medicines and other forms of drugs not considered to be substantial food, and certain continuities occur, as in the cases of nonhuman disease and rectal fistula. The legends recounting specific treatments sanctioned by the Buddha, however, vary among the different versions.[7]

This system of medicine, as formulated and established in the early Buddhist *saṅgha* in India, was transmitted nearly in its original form to other parts of Asia through the activity of Buddhist missionaries and pilgrims.

Extra-Indian Medical Additions to the Common Vinaya Core Evidenced in Recensions of the Jīvaka Legend

The popular Buddhist legend of the physician Jīvaka Komārabhacca, part of a *khandhaka* from the oldest stratum of the Vinaya, provides information concerning the general state of medicine in ancient India and, more importantly, the specific medical lore circulated among Buddhist monks. Additionally, however, comparison of the legend's four recensions—in Pāli, Sanskrit, Tibetan, and Chinese—elucidates a general pattern by which medical knowledge was transmitted and incorporated into the general corpus of Buddhist scripture: a modification of an original story, based on regional and doctrinal influences. The earliest extant version, preserved in Pāli, reflects the current state of medical knowledge and practice in ancient India. The Sanskrit-Tibetan and Chinese accounts indicate accretions based on esoteric Mahāyāna and local influences. In form, the story contains a mixture of factual and fictional material adapted and changed to suit the inclinations of particular Buddhist traditions. The legend's mythical parts are grouped primarily around Jīvaka's early life and the magical and miraculous aspects of his healings, while the factual information is contained mostly in episodes recounting his medical practice.

The legend appears in the chapter on the rules pertaining to the clothing worn by Buddhist monks (Pāli *Cīvarakhandhaka*, Skt. *Cīvaravastu*), which is chapter 5 of the Vinaya of the Mahīśāsaka school, chapter 6 of the Dharmaguptaka, chapter 7 of the Sarvāstivāda and Mūlasarvāstivāda, and chapter 8 of the Theravāda school. The Mahāsāṃghika version is fragmented.[8] Of these, the four recensions examined include the Pāli of the Theravāda school, the Sanskrit and Tibetan of the Mūlasarvāstivāda school, and the Chinese, which is included in a section of various *sūtras* not found in the Vinaya of the schools. The Tibetan version is practically identical to the Sanskrit from which it was translated, and the Pāli has a close variant in Sinhalese.[9] In general, local traditions and doctrinal inclinations were fundamental factors governing the content of the individual versions of the story. The Theravāda (Pāli) version tends to emphasize the pragmatic aspects of Jīvaka's life and career, while the Mahāyāna versions illustrate that the story was adapted to stress magical and miraculous elements in the physician's life.

The opening part of the legend presents the myth of Jīvaka's birth and infancy. The Pāli version began with Sālavatī, a courtesan of Rājagaha, giving birth to a son who was then given to a slave woman, who placed him in a winnowing basket, which was thrown on a rubbish heap. In the Sanskrit-Tibetan account, a promiscuous wife of a merchant from Rājagṛha gave birth to a son of King Bimbisāra, placed the infant in a chest, and ordered maidservants to set the chest at the gate of the king's palace. In the Chinese narrative, a divine virgin named "Daughter-of-the-Mango" (Āmrapāli), who was raised by a Brāhmaṇ, gave birth to a son of King Bimbisāra. The boy was born with a bag of acupuncture needles in his hand and therefore was predestined to become a doctor and a royal physician. His mother wrapped him in white clothes and ordered a slave to take him to the king. This account illustrates influences from indigenous medicine, for the therapeutic technique of acupuncture has been part of traditional Chinese medicine from the first century B.C.E.[10]

The same pattern of modifying the frame story of a healer to fit the interests and concerns of Buddhists in northern India, Tibet, and China characterizes the remainder of the legend. In all versions, the infant is taken and raised by the king's son Abhaya. In the Pāli account, the boy is given the name *Jīvaka* because he was "alive" (from root *jīv*, "to live"), and because a prince cared for him he is called *Komārabhacca* (nourished by a prince). In the Sanskrit-Tibetan account, he was given the name *Jīvaka Kumārabhṛta* (nourished by a prince); and in the Chinese, he is called *K'i-yu* (Jīvaka). The name *Komārabhacca* is, however, probably equivalent to Sanskrit *kaumārabhṛtya*, which, according to traditional *āyurveda*, refers to the branch of medicine involving obstetrics and pediatrics.[11]

Concerning his interest in medicine and his medical education, the different versions exhibit considerable variation. In the Pāli account, Jīvaka, as he approached the age at which he must seek his own livelihood, decided to learn the medical craft. Hearing about a world-famous physician in Taxila, he traveled to that city, famous for education, to apprentice with the eminent doctor. After seven years of medical study, he took a practical examination that tested his knowledge of medicinal herbs, passed with extraordinary success, and, with the blessings of his mentor, went off to practice medicine.

In the Sanskrit-Tibetan version, Jīvaka desired to learn a craft. Seeing white-clad physicians, he decided to become a doctor and studied the art of healing. After acquiring the basics of medicine, he wished to increase his understanding by learning the art of opening skulls from Ātreya, the king of physicians, who lived in the city of Taxila. So Jīvaka went there, took the practical examination on medical herbs and performed other healings, and so deepened his knowledge of medicine that he could even advise his master on therapeutic procedures, thereby earning the latter's respect. This event occurred when Jīvaka, observing Ātreya opening a patient's skull, suggested heating the instrument used to remove small reptiles from the open skull, reasoning that the animals, when touched by the hot instrument, would contract their limbs, thereby facilitating their removal. Ātreya, pleased with Jīvaka's depth of understanding, communicated to him the special technique of opening the skull. Jīvaka eventually left the company of Ātreya and journeyed to the city of Bhadraṅkara in Vidarbha, where he studied the textbook (*śāstra*) called *The Sounds of All Beings* (*sarvabhūtaruta*). During his travels, he purchased a load of wood from a thin and feeble man and discovered in the woodpile of gem (*maṇi*) called "the soothing remedy of all beings" (*sarvabhūtaprasādana*), which, when placed before a patient, illuminated his inside as a lamp lights up a house, revealing the nature of the illness. With this magical diagnostic device, he was then fully prepared to embark on his illustrious career as a physician.

Because Jīvaka in the Chinese version was destined from birth to become a physician, he relinquished all claims to the throne and studied medicine. He found that the education he acquired from local physicians was inadequate and showed their deficiencies in the knowledge presented in the textbooks on plants, medical recipes, acupuncture, and pulse lore, which he had successfully mastered. He therefore instructed them in the essential principles of medicine and gained their respect. Hearing of a famous physician, Ātri (*A-ti-li*) Piṅgala (*Pin-kia-lo*), who lived in Taxila, he traveled to the city to learn medicine from him. After studying medicine for seven years, he took the practical examination on medical herbs and passed it with great success. When Jīvaka departed, his master told him that, although he himself was first among the Indian physicians, after his death, Jīvaka would become his successor. On his travels, Jīvaka encountered a young boy carrying firewood and found he was able to see the inside of the boy's body. Immediately realizing that the bundle of wood must contain a piece of the tree of the Royal Physician (*bhaiṣajyarājan*), who, according to early Mahāyāna scriptures, is a bodhisattva of healing, he bought the wood, discovered a twig of the auspicious tree, and used it to diagnose illnesses in the course of his famous medical practice.

The different versions of Jīvaka's medical education offer significant points of divergence in the transmission of the story. An original popular story existed in Buddhist circles, to which regional variations preserved in Pāli, Sanskrit, Tibetan, and Chinese were added. In general, Jīvaka Komārabhacca exemplifies the mobile, wandering life of the physician in ancient India.[12] The Pāli account lacks any reference to his medical education prior to his study in Taxila with the world-famous physician, who is unnamed. Neither does it recount his desire to learn the special technique of opening the skull and his acquisition of this knowledge, nor does it refer to his study of the textbook on the sounds made by all beings and his obtaining an auspicious gem or piece of wood that allowed him to see the inside of a body like an X ray. Each of these topics deserves further discussion. The names Ātreya in the Sanskrit-Tibetan version and Ātri Piṅgala in the Chinese may refer to Ātreya Punarvasu, the seer (*ṛṣi*) mentioned in the *Caraka Saṃhitā*, who received medical knowledge from Bharadvāja and taught medicine to Agniveśa, among others. Jean Filliozat, however, doubts this equation.[13] Nevertheless, the other accounts strongly suggest that Ātreya is the unnamed physician of the Pāli account. The author of the *Caraka Saṃhitā* mentions two individuals named Ātreya. One, Bhikṣu Ātreya, was a minor figure among the sages who gathered in the Himālayas to receive the knowledge of medicine and held the theory that man and his diseases were products of time (*kāla*).[14] The other, Punarvasu Ātreya, frequently mentioned throughout the corpus, was the major teacher of medicine and was such a significant personality in the medical tradition associated with the *Caraka Saṃhitā* that his name was likely known to the ascetics and physicians connected with the Buddhist community. The venue for Jīvaka's medical training in all versions was

Taxila (Takṣaśilā), a famous site of education in ancient India frequented by Buddhists and Hindus alike and known to the Western world from the time of Alexander of Macedon.

Reference to learning trepanation, the technique of opening the skull, is unique to the Sanskit-Tibetan version. The surgical procedure appears to have been part of an esoteric medical tradition. Moreover, it may reflect an actual practice carried out in a particular geographical area. The archaeological record presents examples of trepanned skulls unearthed from the Neolithic site of Burzahom in Kaśmīr, dating, according to carbon 14, from about 1800 B.C.E., and from Timargarha in northwestern Pakistan, dating from about the ninth to the mid-sixth century B.C.E.[15] Very possibly the religio-medical technique of trepanation was commonly practiced by healers of certain indigenous peoples of northern India and Tibet, and, because of its uniqueness as an esoteric medial practice, was preserved in a legend of an extraordinary physician, contained in sacred literature.

The study of the textbook on sounds and the use of a supernatural gem or medicine tree as a diagnostic device permitting the visual inspection of the body's interior are foreign to Indian medicine and reflect dominant trends of magical medicine in Mahāyāna Buddhism and perhaps of Chinese (i.e., Taoist) medicine. Central to esoteric Buddhism is the use of magical charms (*dhāraṇī*s) or *mantra*s to dispel disease, ward off evils, and secure auspicious outcomes. The textbook on all the sounds made by beings might well be a reference to a collection of potent *dhāraṇīs* and *mantra*s. The gem probaly refers to the marvelous wish-granting *cintāmaṇi* jewel commonly associated textually and iconographically with the bodhisattvas of healing.[16]

The final part of the Jīvaka legend recounts a series of treatment he performed on various people in different places to cure their individual afflictions. The Pāli account of these therapies receives a detailed examination in relation to classial āyurvedic medicine in appendix I. A comparison of the four versions of the cures is our immediate concern here.

The account in the Pāli canon describes six diseases and treatment:

1. A seven-year-old head disease of a merchant's wife from Sāketa, who was cured by nasal therapy using clarified butter and other medicines.
2. A rectal fistula of King Bimbisāra of Magadha, who was cured by the application of ointment with the fingernail.
3. A seven-year-old head disease of a merchant from Rājagaha, whom Jīvaka treated by opening the skull and removing two living creatures.
4. A knot in the bowels caused by acrobatics that was suffered by a

merchant's son from Vārānasī, who was cured by a surgical operation in which Jīvaka untwisted the bowels and restored them to their natural condition.

5. Morbid pallor of King Pajjota of Ujjenī, whom Jīvaka treated and cured with a surreptitious application of clarified butter.

6. A body filled with the "peccent" humors suffered by the Buddha and cured by gentle purgations involving a bath, the inhalation of a purgative dusted on lotuses, and a restricted diet of mild foods.

The Sanskrit-Tibetan account gives thirteen cases of diseases treated and cured by Jīvaka:

1. An itching head wound of a man from Udumbara, diagnosed by Jīvaka with the aid of the magic gem as caused by centipede (*śatapadī*) in the man's head and cured by the technique of skull opening taught by Ātreya.

2. Dropsy of a man from Rohītakaland, who was healed by the intake of radish seeds crushed with a mixture of water and buttermilk (*mūlakabījam udaśvinā*).

3. Twisted bowels (*antrāṇi parāvṛttāni*) of a wrestler from Mathurā, whom Jīvaka diagnosed with the magic gem and revived from death by blowing a powder into the body.

4. A woman from Mathurā whose dead husband was reborn as a worm (*kṛmi*) in her womb (*yoni*) and whose cure involved using fresh meat to lure the worm out of her body.

5. A protruding eyeball of a wrestler from Vaiśālī, whose eye was restored to its proper location by pulling the tendons of his heels, a treatment Jīvaka developed from watching the opening and closing of the eyes and the smiling of the face of a corpse in the Yamunā River as fish twitched its heels.

6. A man from Vaiśālī, into whose ear a centipede had crept and given birth to 700 offspring, cured by beating drums, preparing the ground as if in the summer, and using a piece of raw meat (the same technique used for case 4) to lure the centipedes out of the ear.

7. A Brāhmaṇ from Rājagṛha with an eye disease (*akṣiroga*), for which Jīvaka prescribed a sprinkling of ash (*bhasman*).

8. A man blinded when the Brāhmaṇ prescribed the same treatment for his eye disease and eventually cured by Jīvaka with another remedy, along with the explanation that the elemental nature of the two men was different, requiring a separate treatment for each.

9. A boil (*piṭaka*) on the top of King Bimbisāra's head, which Jīvaka

treated by first bathing the king, next anointing the boil with *āmalaka* (emblic myrobalan or Indian gooseberry) and 500 jugs of water infused with ripening drugs (*pācanīyāni dravyāni*), then opening the ripened boil with a razor (*kṣura*), and finally applying 500 jugs of water steeped with wound-binding drugs (*rohaṇīyāni dravyāni*). When the wound healed and the king recovered, Jīvaka was appointed king of physicians.

10. A householder from Rājagṛha with an internal tumor (*gulma*), whom all the local physicians could not cure. Jīvaka informed the man that remedies were hard to find for his ailment, but the following extraordinary cure, which Jīvaka knew but did not prescribe, ensued: going to the cemetery (*śmaśāna*) to die, the man, hungry, consumed a *babhru*-ichneumon and a *candana*-varan (or monitor lizard) burned in a funeral pyre and drank rainwater found in the cemetery, then drank kodra (*kodrava*) porridge (*odana*) and *mathita* (three parts buttermilk mixed with one part water) in a nearby cow pen. The internal tumor burst open and discharged upward and downward, and the man was healed.

11. Vaidehī, the king's stepmother, who had a boil (*piṭaka*) in the genital region (*guhya*), which Jīvaka treated first by having her sit on a type of poultice (*piṇḍa*) to determine the exact location of the boil, then by applying ripening drugs, surreptitiously lancing the boil by having her sit several times on a poultice containing a concealed knife (*śastra*), then cleaning the opened boil with astringent water (*kaṣāyāmbas*) mixed with wound-binding drugs, and finally applying wound-binding drugs directly to the sore—a healing for which Jīvaka was made king of physicians a second time.

12. An internal tumor (*gulma*) of King Ajātaśatru of Magadha, the son of Bimbisāra, healed by causing the king to believe that he had consumed the flesh of his son Udājibhadra, thereby generating within the king an excessive wrath that made the tumor disappear—a cure earning Jīvaka a third appointment as king of physicians.

13. The Buddha's illness characterized by chills (*śītala*) and runny nose (*abhiṣyandaṃ glānam*), resulting from his constant contact with snow (*hima*) as king of the Himavant Mountains and cured by Jīvaka's ministration of gentle purgations involving the smelling of thirty-two lotus blossoms (*utpala*) infused with purgative drugs (*sraṃsanīyair dravyair*), which resulted in the expulsion of the fourfold "peccant" humors (*doṣas*): those that are loosened but not flowing, those that are flowing but not loosened, those that are both loosened and flowing, and those that are neither loosened nor flowing. Afterward, Jīvaka instructed him to eat yellow myrobalan (*harītakī*) with treacle (*guḍa*) and regularly to consume cream (*maṇḍa*). He did as he was told and was cured.[17]

The narrative legend of Jīvaka does not contain his healing of King Caṇḍa Pradyota's insomnia by a surreptitious application of clarified butter, an episode recounted in the *Vinayakṣudraka* of the Mūlasarvāstivāda Vinaya.[18]

The Chinese version gives six cases attended to and cured by Jīvaka:

1. A twelve-year-old head disease of a woman from Sāketa, whom Jīvaka cured by means of nasal therapy with medicines fried in butter.
2. A knot in the bowels of a noble from Kauśāmbī, healed by an operation on the bowels to untangle the knot and restore the bowels to their natural condition.
3. A head disease of a householder's fifteen-year-old daughter, which caused the girl's death on her wedding day. After learning, from questioning the child's father, that she had grown up with the affliction, Jīvaka used the wood from the magic tree of the royal physicians to diagnose that numerous creatures were eating her brain. He then opened the girl's skull with a golden knife, removed the creatures, closed the skull, and applied three types of supernatural oils to the wound. Seven days later, the young girl was revived and cured.
4. A householder's son from Vaiśālī, who had fallen and died while playing war games. Jīvaka examined the abdomen of the boy with the wood of the magic tree and ascertained that his death was caused by the liver being turned inside out and the obstruction of the vital air. Opening the abdomen with a golden knife, Jīvaka inserted his hand into the cavity and rearranged the liver so that it again functioned properly. He then anointed the wound with the three kinds of supernatural oils. After three days, the boy recovered his full health.
5. Fits of fury of a southern king (Pradyota) who lived a distance of 8000 Chinese *li* from Rājagṛha. The king would often fly into rages and have numerous people executed. During these seizures, he would suffer shortness of breath and suffocation, and his body would burn as if on fire. Debating whether to treat the wicked king, Jīvaka consulted the Buddha, who informed him that in a former life he and Jīvaka had sworn an oath to cure men: the Buddha healed maladies of the soul; Jīvaka, those of the body. After examining the king's pulse and illuminating his body with the magic wood, Jīvaka determined that the poison of a snake had entered his body, causing his viscera, blood vessels, and breath to function improperly. Jīvaka then consulted the queen mother to obtain the composition of the king's remedy, which could only be secretly disclosed and was never to be divulged. In a dream, the queen mother provided

Jīvaka with the vital information that the king was the son of a serpent and that the proper remedy was clarified butter, which the king loathed. Jīvaka decocted the butter so that it did not have the taste of clarified butter, administred it, fled before the king realized what he had taken, and eluded a henchman sent to bring him back. The king eventually recovered, made good all the evil he had done, and rewarded Jīvaka by receiving the Law (*dharma*) from the Buddha, the only form of payment Jīvaka requested for his treatment.

6. The Buddha's illness of cold sweats cured by Jīvaka by the administration of a mild purgative involving the inhalation of a special medicinal dust that falls from a blue lotus flower and the intake of warm water, resulting in thirty purgations. Afterward, the Buddha ate a specific quantity of soft rice, gruel, and broth prepared by Jīvaka and was completely cured.[19]

In this section of the legend, the differing numbers of cures recounted highlight the variations occurring between all the versions. Names of individuals and their locations are changed, and details of the individual therapies are greatly altered, reflecting regional peculiarities, indigenous influences, and doctrinal oddities. All versions contain an account of an operation involving the opening of the skull, or trepanation, to remove pain-causing animals. Similarly, the cure of one suffering from a knot in the bowels or twisted intestines by a type of laparotomy and the healing of King Pajjota (Pradyota) by the surreptitious application of clarified butter are common to the four versions. However, the Sanskrit-Tibetan version places special emphasis on surgery and contains numerous cures reflective of Indian āyurvedic medicine. In both the Sanskrit-Tibetan and the Chinese accounts, an affliction requiring a laparotomy occurs twice, and strong indications of magical medicine under the influence of esoteric Buddhism can be found. Moreover, certain regional and doctrinal influences are highlighted in the Sanskrit-Tibetan account of Jīvaka's treatment of the Buddha. When all versions are compared, variations occur in precise identification of Buddha's affliction. In the Pāli account, it is a humoral disorder; in the Chinese, cold sweats; and in the Sanskrit-Tibetan, a disease characterized by chills and runny nose. When compared with the Pāli, the Sanskrit-Tibetan version mentions the cold climate of the northern mountainous regions as the principal external cause of the Buddha's illness and relates a fourfold division with reference to the "peccant" humors (*doṣa*s), which are the internal causes. This association with cold climate is implied in the Chinese version. The predominance of four rather than three "peccant" humors is characteristic of Mahāyāna medical theory.[20]

A comparative study of the chapter on medicines in Vinaya recensions of the different Buddhist traditions, perhaps enlisting the involvment of various specialists, could provide additional information pertaining to the transmission of medical knowledge via the Buddhist missions and especially the role of medicine in the Buddhist monastic communities of Asia.

Medical Knowledge in Nonmonastic Buddhist Treatises

The combination of two paradigms of healing observed in the Jīvaka story and traced in the early śramaṇic traditions is also evident in sections of later Buddhist literature outside the Vinaya. In the scriptures of the different Buddhist traditions, empirico-rational medicine has an important place, but magico-religious medicine coexists with it on equal footing. Both types of healing are flavored with obvious regional variations. Translations of Indian Buddhist scriptures into various East and Central Asian languages began with Chinese during the second century of the common era. During the seventh century, renderings from Sanskrit into Tibetan began to appear, and between the end of the eighth and the tenth centuries, numerous ranslations from Chinese into Tibetan and vice versa were produced. The greatest production of Khotanese translations also took place during this period. The following survey of medical data in Sanskrit, Khotanese, Tibetan, and Chinese Buddhist treatises illustrates the range of medical practices the Buddhists utilized in their continual involvement with the healing arts and elucidates certain trends in the transmission of Indian medical knowledge by the Buddhist missions.

Sanskrit Sources

The *Suvarṇaprabhāsasūtra* (*The Sutra of Golden Light*) is a Sanskrit Buddhist text containing medical information. Best estimates date the work to the first half of the fourth century C.E. It was translated into Chinese by Dharmakṣema in the fifth century and again by I-tsing in the late seventh or early eighth century. Translations into Tibetan, Uighur, Mongolian, Sogdian, Hsi-hsia (Tangut), and Khotanese also exist. All but the last, whose fragments closely resemble the Sanskrit text, are based on the Chinese translation of I-tsing.[21] Chapter 6 on Sarasvatī, the goddess of speech, offers an explanation of the act of bathing attended by spells and medicines and other incantations, reflecting elements of magico-religious medicine. Chapter 16 on healing diseases adumbrates the principles of āyurvedic medicine. It focuses on the etiology based on the three "peccant" humors (wind, bile, and phlegm), on disturbances occurring during the seasons of the year, and on treatments based on humoral and climatic causes and on an individual's characteristics. This chapter exemplifies the empirico-rational approach to medicine typified in the classical āyurvedic treatises.

The Bower Manuscript is another Sanskrit Buddhist text containing a large section on medicine. The manuscript was discovered in 1890 by Major General H. Bower in Kuchar, a principal oasis and settlement in eastern Turkestān, situated on the caravan route to China. Best estimates place its date from the fourth to the sixth century C.E.[22] Several parts of the manuscript contain medical information; most comprehensive of these is the *Nāvanītaka* ("Derived from Fresh Butter") section, which presents a series of prescriptions along with the diseases they cure. It is essentially a manual of treatments, reflecting a practical rather than an academic approach to medicine.

With the composition of the Sanskrit Buddhist *sūtra Saddharmapuṇḍarīka* (*The Lotus of the True Law*), whose first twenty-two chapters probably existed before 100 C.E., there emerges the figure of the bodhisattva of healing—Bhaiṣajyarājan (The King of Healing) and Bhaiṣajyasamudgata (Supreme Healer).[23] Both would subsequently have illustrious careers in China. By reciting their names or the *dhāraṇīs* they pronounced, devotees would be assured of health, long life, and numerous benefits. Bhaiṣajyarājan gave rise to the healing bodhisattva par excellence, Bhaiṣajyaguru (The Teacher of Healing), who appears in numerous Mahāyāna works and who has an entire *sūtra* devoted to him (*Bhaiṣajyagurusūtra*). He eventually attained the full rank of a Buddha with a large following of devotees in the Mahāyāna Buddhism of East Asia. The bodhisattva of healing in the *Saddharmapuṇḍarīka*, Bhaiṣajyarājan, performs the principal function of propounding the Buddhist doctrine (*Dharma*) and was specifically singled out by the Buddha for this purpose. His extremely compassionate nature is recounted in the tale of his self-immolation in a past life (Chapter 22), exemplifying the selfless compassion required of a healer. Later developments of the notion of a healing buddha, especially in East Asia, present as his major function the healing of mental defaults that prevent spiritual growth and inhibit enlightenment, and helping others to overcome these handicaps; yet he is aiso supplicated in order to remove and heal physical afflictions.

The healing Buddha Bhaiṣajyaguru, seemingly linked to the worship of Bhaiṣajyarājan, appears to have originated or at least become the object of cult worship in Central Asia or Kaśmīr during the third century C.E. and was already important in China in the fourth century, although the *sūtra* devoted to him achieved popularity in Central Asia around the seventh century C.E.[24] His twelve vows, recorded in the *sūtra*, exemplify the general functions of the Buddhas and bodhisattvas of healing and were to become operational on his attaining Buddhahood:

1. May a world be illuminated by the rays of my body and may all beings be endowed like me with the signs of the great man (*mahāpuruṣa*).
2. May my body be so resplendent as to surpass even the brightness of the sun and the moon and make dark nights bright enabling beings to move about easily.
3. May my infinite knowledge and acquisitions offer protection and help to the beings and in consequence may there not be any deformed beings.
4. May all beings take to Mahāyāna, leaving aside false doctrines and Hīnayāna.
5. May all those joining the order be self-restrained and observant of the precepts and may they not be born in evil states after hearing my name.
6. May every being be cured of his deformity on hearing my name.
7. May every ailing being too poor to afford medicine be cured of his maladies on hearing my name.
8. May all feminine beings get rid of their femininity on uttering my name.
9. May all beings be turned by me from false to right views and ultimately to bodisattva practices.
10. May all those destined to be punished by the king be relieved of their sufferings on hearing my name.
11. May the famished transgressing even the law for the sake of food obtain excellent food on hearing my name.
12. May all those destitute of clothing obtain attractive clothes on uttering my name.[25]

The bodhisattva of healing is unique to Mahāyāna Buddhism and, according to David Snellgrove, is "the special manifestation of buddha-hood responsible for the Indian medical traditions, which were exported as a useful part of Buddhist culture in general."[26] The idea of spiritual healing and the practice of curing by the propitiation of a deity were assimilated and adapted from medical ideas current in northwestern India, Central Asia, and China, where indigenous forms of magico-religious medicine predominated.

Khotanese Sources

Several medical treatises occur in the Central Asian language of Khotanese. Most are fragmented, but two are complete and have Sanskrit equivalents from which the Khotanese was presumably translated.[27] Ravigupta's *Siddhasāra*, dating from about 650 C.E., has extant versions in Sanskrit,

Khotanese, Tibetan, and Uighur. It is a complete medical manual based on the theories and principles of *āyurveda*.[28] The second undated text, labeled *Jīvakapuṣṭaka*, has a corrupt Sanskrit version and is a concise recipe book of medical therapies for different ailments, closely resembling in style the medical portions of the Bower Manuscript.[29]

Tibetan Sources

Tibetan Buddhism, like Indian Buddhism, had a great interest in medicine, especially Indian medicine, and faithfully transmitted Indian medical knowledge to Tibet. Indian Buddhist scholars responsible for conveying Buddhist knowledge to Tibet were conversant with Indian medical doctrines. This is perhaps best illustrated in a possible eighth-century C.E. letter to the Tibetan King Khri Srong-lde-brtsan by the Indian Buddhaguhya, renowned in Tibet as an authority on esoteric Buddhism, in which, in addition to expounding Buddhist Tantric doctrines, the Indian savant explains that his ill health, caused by the "combined disorders of wind and bile and phlegm," prevents him from making the journey to Tibet.[30] Accumulated Indian medical knowledge appears in sections of the Tibetan Buddhist canon devoted to medicine (*gso-rig-pa*). They contain numerous Tibetan translations of Sanskrit medical treatises and commentaries, most of which are extant and exemplify the empirico-rational medicine of *āyurveda*.[31] Two texts from this collection deserve special mention. Vāgbhaṭa's *Aṣṭāṅgahṛdayasaṃhitā* (Tibetan *Yan-lag brgyad-paḥi sñiṅ-po bsdus-pa shes-bya-pa*) (*The Collection of the Essence of the Eight Limbs* [*of Āyurveda*]), dating most probably from around the seventh century C.E., is generally acknowledged to be the first compilation of medical knowledge contained in the early treatises of classical Indian medicine. It expounds both theory and practice and seems best suited for the academic study undertaken in larger Buddhist monasteries. It still receives wide acclaim among the Nambudiri Brāhmaṇs of the Aṣṭavaidya medical tradition in Kerala.[32] Nāgārjuna's *Yogaśataka* (Tibetan *Sbyor-ba-brgya-pa*) may date from the seventh century, although Filliozat improbably attributes its authorship to the Mādhyamika Buddhist savant Nāgārjuna, who lived in the second century C.E.[33] It is essentially a manual of practical medicine, similar to the *Nāvanītaka* of the Bower Manuscript.

In Tibet and Mongolia, the principal medical treatise, called the *Rgyud-bźi* (*Fourfold Tantra*), is attributed to Bhaiṣajyaguru and does not seem, as previously thought, to be a translation from a lost Sanskrit original of the eighth century C.E. It is a composite work incorporating elements of Chinese medicine within the framework of Indian medicine. We find

Chinese names for plants whose properties are described according to the Indian system and Indian names for bodily organs classified into five solid and six hollow organs precisely according to the Chinese system of classification of yin and yang organs. Reference is made to Indian medical etiology, but also the Chinese techniques of examination of the pulse and urine also are used to determine the causes of disease.[34]

Other Tibetan Buddhist treatises preserve esoteric Mahāyāna doctrines of magico-religious medicine, including the reverence of the healing Buddhas and bodhisattvas, the use of the *dhāraṇīs*, and magical healing rituals. Much of the medicine in Tibetan Buddhist works combines medical knowledge derived from both India and China, largely through Tibetan translations undertaken from both Sanskrit and Chinese sources. It is probable that elements of esoteric Buddhism and of indigenous (perhaps Bonpo) traditions also contributed much to the medicine exemplified in Tibetan Buddhist texts.

Chinese Sources

Buddhist treatises containing medical information found in the Chinese canon are mostly translations of Sanskrit sources no longer extant. As far as can be ascertained from the available data, Chinese Buddhist literature does not contain translations of Indian medical treatises. For the most part, medical information derives from original Sanskrit Buddhist *sūtras* and from the texts devoted to the healing Buddhas and bodhisattvas. Specific medical knowledge preserved in the Chinese texts is based largely on the theories and practices of Indian medicine brought to China by Buddhist pilgrims beginning from the fourth century C.E. and incorporates elements of both magical medicine and indigenous Chinese medical lore.

The followers of Mahāyāna doctrines used medicine as a vehicle of conversion. Paul Demiéville has excellently summarized the medical data in the Chinese Buddhist sources in his article "Byô" in the third fascicule of the *Hôbôgirin*, an encyclopedic dictionary of Buddhism.[35] The following information derives from his essay.

Three trends stand out as fundamental to the medical principles of Chinese Mahāyāna Buddhism: (1) the superimposition of an etiology based on four elements on a newly conceived fourfold theory of disease causation derived from the traditional threefold humoral theory (*tridoṣa*) characteristic of āyurvedic medicine; (2) a threefold classification of treatments that incorporate traditional Chinese medical techniques of acupuncture, cauterization (moxa), and pulse lore; and (3) the role of Buddhas and bodhisattvas as healers.

Buddhist literature in general and Chinese Mahāyāna texts in particular ascribe the origin of disease to a disequilibrium of four elements (*mahābhūta*s or *dhātu*s) of which the human body as well as all matter is constituted: solid-earth, wet-water, hot-fire, and mobile-wind. The total number of diseases is usually given as 404, or 101 for each of the four elements. Evidence of an eightfold classification of disease causation occurs in the Pāli canon. Of these eight, four are attributed to the three "peccant" humors (*doṣa*s) (wind, bile, and phlegm) plus a fourth, understood to be a combination (*sannipāta*) of the three.[36] A traditional Chinese conception of etiology, based on the principal elements, was superimposed on these four. This homologization of two medical ideas resulted a medical theory intelligible to the Chinese. Mahāyāna Buddhists also fit the notion of the Four Noble Truths and the four qualities of a great physician into this fourfold system.[37]

The general classification of treatments into three types derives from data contained in various *sūtra*s devoted to the curing of numerous afflictions. The three forms of medical therapeutics were (1) religious healing, divided into external types, involving confession, and internal ones, comprising, among others, mental exercises, meditation, and insight; (2) magical healing, involving, among others, incantations and exorcisms; and (3) proper medical healing, including, among others, the use of drugs, dietetics, and surgery. Therapeutic techniques described in the Chinese sources often incorporated aspects of classical Chinese medicine, especially techniques of magical and religious healing. Cures involving proper medical treatment often included the use of acupuncture needles, cauterization by means of moxa, and examination of the pulse. None of these three forms of traditional Chinese medical therapeutics is found in the medical passages of the Pāli canon or in the texts of classical āyurvedic medicine.

Whenever the eight limbs (*aṣṭāṅga*) of Indian medicine are mentioned, the branch called "major surgery" is defined as the use of needles, especially in the treatment of cataracts and eye diseases. Indian cataract surgery seems to have been introduced into China between the seventh and ninth centuries C.E. Medical treatment and surgical operations on the eyes were therapeutic measures for which Indians acquired a great reputation both in China and in Greece.[38] Eye care was well known in early Buddhist monastic medicine as well as early *āyurveda*, and an example of eye surgery occurs in the Pāli Sivijātaka, depicted in Buddhist art from Bhārhut.[39]

The *Rāvaṇaproktabālācikitsāsūtra* (*Sūtra Spoken by Rāvaṇa on the Curing of Children's Diseases*) is a tenth-century Chinese translation of the Sanskrit treatise *Rāvaṇakumāratantra*. A work on pediatrics, it treats diseases of demonic possession in children to age twelve. The treatments are magical, involving offerings to demons, purificatory baths, fumigations, recitations of magico-religious utterances (*mantra*s), and other works of piety. Contrary to Filliozat, P. C. Bagchi claims that it is a work of Buddhist inspiration and states that "the Chinese text seems to have preserved the most correct form of the Rāvaṇakumāratantra."[40]

Demonstrating a close affinity with the *Rāvaṇakumāratantra* is the *Kāśyaparṣiproktastrīcikitsāsūtra* (*Sūtra of Gynecology Taught by the Ṛṣi Kāśyapa*), a fragmented tenth-century Chinese translation of a Buddhist treatise on embryology and prenatal care said to have been taught to the *ṛṣi*'s apprentice Jīvaka. A short, strictly medical work prescribing medication for each of the ten months of pregnancy, it is closely connected to the *Kāśyapasaṃhitā*, an incomplete āyurvedic text devoted to gynecology and children's diseases, which is ascribed to a certain Vṛddhajīvaka ("Jīvaka the Elder") and which in turn is based on the earlier treatises of the *Caraka* and *Bhela Saṃhitā*s.[41] When the relevant sections (*jātisūtrīya*) in the four texts are compared, the Buddhist *Sūtra* emerges as a clear derivation of the *Kāśyapasaṃhitā*.[42] This connection affords further evidence of links between Buddhist medicine and *āyurveda*, which found their origins in the tradition of early śramaṇic physicians.[43]

In addition to the *Kumāratantra* of Rāvaṇa, numerous texts involving magico-religious healing are found in the Chinese Buddhist canon. These *sūtra*s were uttered to cure numerous ailments, including toothache, hemorrhoids, eye disease, and childhood diseases. They complement the magical medicine surrounding the cult of the Buddhas and bodhisattvas of healing by attracting followers who seek out Buddhists possessing the powerful magical utterance to have their various ills removed.[44]

As noted in Chapter 3, early Buddhist monks learned practical aspects of medicine as part of their monastic training for the specific purpose of providing care for their brethren. In later forms of Buddhism, however, monks were permitted and even encouraged to learn both the theoretical and the practical aspects of medicine, included among the five sciences (*vidyā*) in the major monasteries. Medicine was considered one of the means (*upāya*) to liberation and a subject that a bodhisattva, either monk or lay, was obliged to acquire. Moreover, the obligation to heal the sick is stipulated in the disciplinary codes of Mahāyāna Buddhism. This democratization of medicine was part of the new movement in Buddhism characterized by Mahāyāna. By learning the appropriate magical utterances and their accompanying rituals, as well as transmitted techniques of empirico-rational āyurvedic medicine, the followers of the bodhisattva path could heal both spiritual (mental) and physical afflictions. The magical potency believed to be contained in the name of the healing Buddha or bodhisattva corresponded to the Mahāyāna notion of magical charms (*dhāraṇīs*). In this way, the recitation of the Buddha's or bodhisattva's name was thought to be an efficacious cure. It was explicitly understood that the healing of the body permitted the calming of the mind and the cultivation of awakening.

The notion of the healing bodhisattva, as previously mentioned, began in early Mahāyāna and seems to have developed and spread from northwestern India or Central Asia to China, where the Buddhas and bodhisattvas of healing became significant aspects of East Asian Buddhism. Entire cults, such as that devoted to Bhaiṣajyaguru, evolved, and an iconography of the deity was established. The worship and ritual surrounding the Buddhas and bodhisattvas of healing reflect a strong magico-religious element, characteristic of Mahāyāna Buddhism in Central and East Asia. In China, rituals of propitiation and worship of healing Buddhas and bodhisattvas were expected to secure from them physical, mental, and spiritual healing, as well as spiritual guidance. Such rituals ranged from a simple *pūjā* to elaborate, complex ceremonies involving the recitation of magical utterances, including the names of the Buddha and bodhisattva, and specific actions around sacred *maṇḍala*s and their images. The iconography of the Buddha Bhaiṣajyaguru included the magically potent stone lapis lazuli, symbolizing purity and rarity, and yellow myrobalan (*harītakī*), a well-known healing substance found both in Buddhist medicine and in *āyurveda*.[45] The image combines both the magico-religious healing, exemplified in the medical traditions of Central and East Asia, and the empirico-rational medicine, typified by the āyurvedic medical tradition of India. The deity, therefore, symbolizes a perfect healing system, uniting the medical lore of all the Buddhist traditions from India to China.

From the available textual evidence, it is clear that Indian medicine, as represented in the Buddhist treatises, was not accepted in China in the same way as it was in Central Asia and Tibet, where it was assimilated nearly verbatim. Perhaps being considered too foreign for the refined Chinese, Indian medicine had to be modified to fit the Chinese outlook. For example, the Indian etiological conception of disease causation based on the idea of "peccant" humors (*doṣas*) appeared in Chinese sources only in the form of a fourfold system based on the principal elements (earth, air, fire, and water), of which the Chinese had a much clearer conception, even though they believed there existed a fifth element, ether. Actual medical therapeutics emphasized indigenous Chinese techniques. The magico-religious aspects of esoteric and Mahāyāna Buddhism, including the tradition of the Buddhas and bodhisattvas of healing, were especially well received in China, which already had an established tradition of magical medicine associated with Taoism. The elements of Buddhist medicine from India were so greatly altered and adapted by the Chinese Buddhists that it becomes difficult to recognize them.

Evidence from the literary sources preserved in the Buddhist traditions of India, Central Asia, Tibet, and China presents definite trends in the spread of Indian medical knowledge. Documents setting forth the monastic rules pertaining to medicine and healing were faithfully transmitted to various countries of Asia with the early Buddhist missionaries. However, the legend of the physician Jīvaka, recounted in the monastic texts, illustrates the infiltration of indigenous and magical medical ideas.

The influence of sectarian doctrines and indigenous medicine is most widely observed in the medical material contained in nonmonastic documents of later Buddhism. A significant influence of magico-religious forms of healing dominated much of the medicine of these later traditions. This is particularly apparent in the esoteric Buddhist doctrines and in rituals and practices surrounding the cult of the buddhas and bodhisattvas of healing in China. The Sanskrit medical documents of Mahāyāna preserve much of classical Indian medical knowledge. At the same time, they propound magical techniques of healing in the form of *dhāraṇī*s, *mantra*s, incantations, and the use of healing herbs. These texts contain the beginnings of the notion of the bodhisattva of healing. Tibetan Buddhist literature also preserves a great deal of Indian medical knowledge, including translations of important Sanskrit medical treatises; yet evidence of esoteric forms of healing and the cult of the Buddha of healing can also be found. In addition, Tibetan medicine, through the influence of Buddhism, developed its own body of literature that demonstrates a mixture of Indian, indigenous, and Chinese medical lore. The Chinese Buddhist sources reflect Mahāyāna doctrinal ideology. The cult of the healing Buddhas and bodhisattvas reached its full development in China. Likewise, Indian medical theories were changed and adapted to fit Chinese conceptions, and traditional Chinese healing was integrated into Buddhist doctrine and practice. Magico-religious healing was accepted by Chinese Buddhists, although very little evidence of original Indian medical ideas and practices remained in the Chinese Buddhist treatises.

As Indian medical knowledge and practice spread with the diffusion of Buddhism, it underwent transformations and adaptations. The medical lore of the original Buddhists was preserved virtually unchanged, was modified, or was wholly discarded and replaced with more suitable ideas and practices. The fundamental factors contributing to the particular course of development that medicine took among the different Buddhist traditions derived largely from specific regional and doctrinal influences. One important aspect remained constant, however: medicine was always a significant part of Buddhism throughout the development of the religion.

II

THE CONTENT OF EARLY BUDDHIST MONASTIC MEDICINE

Part II embarks on a philological investigation of the medicines and medical therapies sanctioned in the monastic code (Vinaya) of the Pāli canon, beginning with the enumeration of the materia medica of the religious community and concluding with the therapeutics employed for specific ailments. The section on medicines in the *Mahāvagga* represents the earliest codification of medical knowledge in India, is characterized by reference to actual cases, and functioned as a handbook and guide for the treatment of common afflictions. Authority for inclusion of those medicines and remedies in the monastic code was typically ascribed to the Buddha by the technique of representing cases requiring medicines and therapies as being presented to him for the sanctioning of remedies. It is uncertain whether all the reported cases actually came to the Buddha's attention or are merely examples of this compilatory technique. The medical importance of these Buddhist records, however, is their recounting of patients' medical problems and the corresponding treatments. The academic medical treatises of classical *āyurveda* offered no such case-by-case medical instruction.

Through a careful study of medical material in the Buddhist Pāli records, a clear picture of Buddhist monastic medicine emerges and, when compared with the relevant sections of the classical medical treatises of the *Caraka*, *Bhela*, and *Suśruta Saṃhitā*s, provides a deeper understanding of the common storehouse of śramaṇic medicine from which the Buddhists and compilers of the medical treatises derived their respective medical data. Between those two traditions many similarities, but also numerous differences, exist—significantly, the Buddhist emphasis on practical application devoid of the theoretical considerations of disease etiology that dominate the medical books. This difference supports the view that codified Buddhist monastic medicine, with its emphasis on materia medica and case-based therapies, represents an early attempt to provide a manual of medical practice and in some sense legitimated the formalized collections of prescriptions detailed in later Buddhist medical recipe books and enchiridions such as the Bower Manuscript's *Nāvanītaka*, Nāgārjuna's *Yogaśataka*, Ravigupta's *Siddhasāra*, and the *Jīvakapuṣṭaka*.

5

Materia Medica

The materia medica of the Buddhist monastery, representing the medicines requisite in sickness (*gilānapaccayabhesajja*), included initially the five basic medicines (ghee, or clarified butter; fresh butter; oil; honey; and molasses, or treacle), to which were added a more extensive pharmacopoeia of fats, roots, extracts, leaves, fruits, gums or resins, and salts. Medicines of the Buddhist monastic materia medica were considered to be foods but classified as nonsubstantial nourishment, allowing them to be consumed at any time. The monastic code prohibits monks and nuns from eating food between midday and sunrise.[1] Medicinal foods are exceptions to this rule. Certain foods were classified as medicines not merely to allow Buddhist cenobites to eat at any time, but because these provisions afforded considerable benefit to the monastic community. The length of time each medicinal food may be stored before spoiling is specifically mentioned, and the items listed presumably represented only the most commonly used and most readily available drugs because the code ordinarily stipulates that any medicinal food can be used as medicine as long as it is not substantial nutrition. This stipulation implies that the actual pharmacopoeia was more extensive than the list of specific drugs would indicate. Classification of particular foods as medicines and specification of storage periods point to a derivation of the Buddhist monastic materia medica from ancient Indian culinary traditions, which, likewise, appear to have been the basis of aspects of early āyurvedic pharmacopoeias because many codifications of drugs in the medical treatises appear in the sections related to food and drink (*annapāna*).

Five Basic Medicines

The five basic medicines are introduced in an account relating how certain monks at Sāvatthi suffered from a disease that occurred during the autumn months (mid-September to mid-November). The symptoms for the disease are explicit: the monks threw up the rice gruel they drank and the food they ate. Thereupon, they became emaciated, miserable, of bad color, and increasingly pale (or jaundiced), and their limbs were covered with blood vessels (*dhammani*). The Buddha was consulted and sanctioned five medicines (*bhesajja*), deemed to be food but not ample nourishment, to be given to the monks: ghee, or clarified butter (*sappi*, Skt. *sarpis*); fresh butter (*navanīta*); oil (*tela*, Skt. *taila*); honey (*madhu*); and molasses, or treacle (*phāṇita*). After taking the medicines, the monks did not improve, and their symptoms became increasingly acute, for they were unable to digest ordinary coarse meals or even less oily ones. As a result of this worsening condition, the monks were permitted to partake of the five medicines at any time (i.e., before or after the noon hour).[2]

The fifth-century C.E. commentator Buddhaghosa explains that this disease is an affliction of the "('peccant') humor bile" (*pitta*), arising in the autumn. "At that time," he says, "they [i.e., the monks] become wet by rainwater and trample in the mud. The heat occasionally becomes painful. Therefore, their bile becomes situated in the intestinal canal [*koṭṭhabbhantara*]." He adds that the complication of indigestion is attributable to the vitiation of wind (*vāta*).[3]

In another part of the monastic code, these five basic medicinal foods, which can be stored for a maximum of seven days, are defined in relation to their sources:

> "Clarified butter" is clarified butter from [the milk of] cows, she-goats or buffaloes, that is, the clarified butter of those [animals] whose meat is suitable; "fresh butter" is made from the milk of those same [animals]; "oil" is the oil of sesame seeds [*tila*, Skt. *tila*] or mustard seeds [*sāsapa*, Skt. *sarṣapa*], from the mahua plant [*madhuka*, Skt. *madhūka*][4] or the castor plant [*eraṇḍa*, Skt. *eraṇḍa*], and from animal fats; "honey" is from bees; and "molasses" is produced from sugarcane.[5]

Examination of these medicinal foods in the medical treatises of Caraka and Suśruta reveals a concordance between the two medical traditions, indicating a common origin. The five medicines with similar sources are codified in Suśruta's chapter on fluid substances (*dravadravya*) and Caraka's chapter on various foods and drinks (*annapāna*):

1. Clarified butter (*ghṛta*) occurs in Suśruta's group of clarified butter

(*ghṛta*) and Caraka's group of cow's products (*gorasa*), alleviates wind and bile, and is prescribed especially in the autumn.[6]

2. Fresh butter (*navanīta*) occurs in Suśruta's buttermilk group (*takra*) and Caraka's group of cow's products and removes vitiated bile and wind.[7]
3. Oil (*taila*) occurs in Suśruta's group of oils (*taila*) and Caraka's group of food additives (*āhārayogin*), promotes digestion, and appeases wind and phlegm.[8] Oil from sesame seeds (*tila*), the best oil, removes wind and promotes appetite.[9]
4. Honey (*madhu*) is found in Suśruta's honey group (*madhu*) and Caraka's sugarcane (*ikṣu*) group, aggravates wind, and alleviates bile and phlegm.[10]
5. Molasses, or treacle (*phāṇita*), occurs in Suśruta's sugarcane group (*ikṣu*), is nourishing, and increases all three humors. According to Ḍalhaṇa, the twelfth-century C.E. commentator on the *Suśruta Saṃhitā*, it is also called minor jaggery (*kṣudraguda*) or treacle, which is mentioned by Caraka in his sugarcane group as a producer of fat and muscle.[11]

The early medical treatises also provide correspondences to the specific affliction suffered by the monks. Both Caraka and Suśruta teach that the bile (*pitta*), accumulated in a body that has become accustomed to the cooling rains of the previous season, commonly becomes aggravated after the body is exposed to the heat of the sun's rays in autumn. In order to counteract the bile, substances that pacify that humor are recommended. These include sweet, light, cool, and slightly bitter foods used in proper quantity by those with good appetite. Oil (*taila*) and fat (*vasā*) are to be avoided.[12]

The disease suffered by the monks is described in the medical treatises as caused by the excitement of bile during the autumn. The Buddhist account offers no indication that the affliction involved bile. This information is supplied by Buddhaghosa. The prescribed Buddhist remedy was the five medicinal foods, three of which remove bile (i.e., clarified butter, fresh butter, and honey) and one of which exacerbates bile (i.e., molasses), indicating that only those medicinal provisions effective against the precise disease were used. The permitted treatment of the complication of indigestion, which Buddhaghosa attributed to wind, included these five medicinal foods necessary to alleviate that morbid condition (probably clarified butter, oil, and perhaps also molasses). The proper and allowable treatment for the disease suffered by the Buddhist monks, therefore, required the use of all five medicinal food provisions. There is clear continuity between Buddhist monastic and early āyurvedic medicine, indicating the derivation of medical knowledge from a common source of medical lore. The Buddhist's detailed description of the illness's observable manifestations illustrates the emphasis on practical rather than theoretical concerns and does not imply the absence of humoral etiology at this time. Buddhaghosa's later supplementation of humoral-based causes in his commentary indicates the Buddhist savant's familiarity with āyurvedic medical theory and practice.

The monastic code follows the enumeration of the five medicinal provisions with the remaining foods constituting the Buddhist monastic pharmacopoeia. Unlike the previous case, a standardized formula introduces the individual groups of medicines: sick monks required a particular medicinal food, in response to which the Buddha was consulted and permitted the food to be used as medicine, thereby increasing the Buddhist monastery's catalog of beneficial foods. Seven additional items, beginning with fats and ending with salts, constitute the medicinal foods of Buddhist monastic materia medica.

Fats

Sick monks (*gilāna bhikkhu*) required fats (*vasā*) as medicines. The Buddha allowed the following five types of animal fat to be used as medicines and consumed with oil (*tela*) at the proper time (i.e., before meals):[13] fats from bears (*accha*, Skt. *ṛkṣa*), from fish (*maccha*, Skt. *matsya*), from alligators (*susukā*), from swine (*sūkara*, Skt. *śūkara*, *sūkara*), and from donkeys (*gadrabha*).[14] In his comments to the Nissaggiya (7.2), Buddhaghosa elaborates on fats as medicines. He says that fat from the flesh of all edible animals and of the ten inedible animals (i.e., man, elephant, horse, dog, snake, lion, tiger, leopard, bear [!], and hyena), with the exception of man, is permitted (as medicine). He states that it is the oil made from the fats, or the fatty oil (*vasātela*)—that is, the oil of the permitted fats (*anuññātavasānaṃ tela*)—that is to be used as medicine.[15]

Although the early medical texts present no clear classification of fats (*vasā*) based on their sources, the authors incorporate fat along with marrow and flesh in their chapters on medicinal foods and list the different animals from which fat is obtained. Caraka includes fat and marrow under the category of food additives, and he states that the two are sweet, nourishing, aphrodisiac, and strength giving, and enumerates fish, birds and wild beasts, crocodile, tortoise, porpoise, boar, buffalo, and ram among animals yielding beneficial fat.[16] Suśruta incorporates fat along with marrow under the category of oils, and generalizes its sources to include domesticated animals and swamp-dwelling animals, whose fat appeases wind; jungle animals, carnivores, and animals with single (nonsplit) hooves, whose fat cures hemorrhagic disorders (*raktapitta*); and birds, whose fat removes phlegm.[17] The medical tradition's codification system illustrates what Francis Zimmermann identifies as a typical brāhmaṇic classification of animals and their products according to a fundamental ecological division between the wetlands (*anūpa*) and the drylands (*jāṅgala*).[18] In recipes requiring the use of fat, oil (*taila*) or ghee (*ghṛta*) is often mentioned.[19]

The Buddhist monastic medical tradition established a separate category of medicinal fats based on their animal sources. The medical authors offer no such separate codification of fats as foods but consider fats together with marrow as a type of food and incorporate fat in a general list of animal products. Their method of classifying fats according to animal types and their habitats illustrates a brāhmaṇic systemization superimposed on an established medical tradition, common to Buddhist monastic medicine, which identified fats as medicinal foods.

Roots

Sick monks needed roots (*mūla*) as medicines. After being consulted, the Buddha permitted the following medicinal roots: turmeric (*haliddā*, Skt. *haridrā*), ginger (*siṅgivera*, Skt. *śṛṅgavera*), sweet flag or orrisroot (*vaca*, Skt. *vacā*), white variety of sweet flag (?) (*vacattha*),[20] Indian atees (*ativisa*, Skt. *ativiṣā*), black hellebore (*kaṭukarohiṇi*, Skt. *kaṭurohiṇī*), vetiver (*usīra*, Skt. *uśīra*), and nut grass (*bhaddamuttaka*, Skt. *bhadramusta*). They can be stored for a lifetime (i.e., indefinitely) and must never serve as solid or soft food but are to be added to substantial nutrition.[21] Horner, without citing supporting evidence, states that the roots mentioned here "are allowed medicinally for flavouring foods which otherwise would be unpalatable for ill monks to take. Decoction of these roots is used today in Ceylon as medicine for fever and stomach complaints."[22]

The Buddha also allowed two types of grindstone for preparing the medicinal roots.[23] According to Buddhaghosa, one stone was large, and the other small.[24] The roots were probably placed in a depression in the larger stone or on top of its flat surface and crushed and pounded into a powder or a pap with the smaller.[25]

The early medical treatises provide a detailed classification of roots (*mūla*) from numerous vegetal sources, for medical authors considered the root as the most medicinal part of a plant. Caraka enumerates sixteen plants that have useful roots,[26] none of which occurs in the Pāli list. He and Suśruta more elaborately divide the roots into five groups of five each: the small (*kanīyas*) group of five roots that eliminate wind and appease bile; the large (*mahant*) group of five roots that remove phlegm; together, these first two are known as the ten roots (*daśamūla*); the groups of five roots of creepers (*vallī*) and of thorny scrubs (*kaṇṭhaka*), which destroy phlegm; and the group of five roots of grasses (*tṛṇa*), which pacify bile.[27] None of the roots enumerated in these five groups appears in the Buddhist list of medicinal roots. An elaborate classification is wanting in Bhela's medical compendium, but various kinds of roots are mentioned along with other usuable parts of plants throughout the entire work.

The Buddhist canonical and post canonical presentation of medicinal roots is considerably more abbreviated than that found in the medical treatises, and the Buddhist sources offer a classification of roots simpler than that found in the medical works. Nevertheless, both sources recognize roots (*mūla*) as important medicines and have included them in their respective pharmacopoeias.

Extracts

Sick monks needed extracts (*kasāva*, Skt. *kaṣāya*) as medicines. The Buddha allowed the following medicinal extracts: those made from the Indian lilac or neem tree (*nimba*), from the kurchi tree (*kuṭaja*), from the *pakkava*,[28] and from the Indian beech (*nattamāla*, Skt. *naktamāla*). Likewise, monks were permitted any extract that does not serve as solid or soft food. The extracts can be stored indefinitely.[29]

Although extracts suggest derivative drugs rather than foods, they seem in the first instance to have been considered as types of food or additives, since the text states that they cannot be used as solid or soft food.

In the early medical texts, *kaṣāya* has a twofold meaning. On the one hand, it delineates astringent as one of the six tastes (*rasa*) and, on the other, it indicates an extract or a decoction prepared from drugs possessing five of the six tastes, including astringent (saline is eliminated because it cannot be processed into other forms).[30] The five kinds of extracts are enumerated in the *Caraka Saṃhitā*: juice (*svarasa*), paste (*kalka*), decoction (*śṛta*), cold infusion (*śīta*), and hot infusion (*phāṇṭa*).[31] Caraka goes on to explain the fifty great extracts, of which the group of ten antipruritics (*kaṇḍūghna*) includes three of the four plants mentioned in the Pāli: *naktamāla*, *nimba*, and *kuṭaja*.[32] Two of the four are found in Suśruta's *lākṣā* group (i.e., *kuṭaja* and *nimba*), the contents of which are said to consist of astringent, bitter, and sweet tastes.[33]

The theoretical understanding of an astringent as a taste expressed by the medical authors is wanting in the Pāli sources. Rather, the Buddhists enumerated four drugs serving as the principal sources of medicinal extracts. The medical tradition's explanation of *kaṣāya* as extract corresponds to the Buddhist treatment of extracts, and a near-exact enumeration of the Pāli list is in Caraka's section of antipruritics. A category of extracts occurs in both early Buddhist and early āyurvedic materia medica.

Leaves

Sick monks required leaves (*paṇṇa*, Skt. *parṇa*) as medicines. The Buddha permitted the following medicinal leaves: the leaves of the Indian lilac (*nimba*), of the kurchi tree (*kuṭaja*), of the wild snake gourd (*paṭola*), of the holy basil (*sulasī* [*ā*], Skt. *surasī* [*ā*]), and of the cotton tree (*kappāsika*, Skt. *kā̆rpāsikā*). Similarly, any leaves that do not serve as solid or soft food may be used; the leaves can be stored indefinitely.[34]

The early Sanskrit medical treatises offer no individual category of leaves (*parṇa*, *patra*) in their general classification of medicines. Caraka mentions leaves in his general enumeration of plant growths (*audbhida*) but does not list them in any fashion.[35] Similarly, leaves are enumerated along with sap, roots, barks, flowers, and fruits as the sources of Caraka's 600 evacuative drugs.[36] The closest approximation to a classification of medicinal leaves is Suśruta's discussion and list of the leaves of various types of vegetables or potherbs (*śāka*) in his chapter on foods and drinks.[37] None of the plants from the Pāli list is mentioned. Elsewhere, certain leaves are included as ingredients of different types of remedies.[38]

The Buddhists understood leaves (*paṇṇa*, *patta*), like extracts, to be a separate group or category of medicinal foods and enumerated their most important types. The early medical tradition had no such classificatory understanding of leaves but included them as a principal sources of drugs used in numerous medicinal therapies. Clearly, both traditions considered leaves as essential components of their respective pharmacopoeias.

Fruits

Sick monks needed fruits (*phala*) as medicines. The Buddha allowed the following medicinal fruits: the fruits of the embelia (*vilaṅga*, Skt. *viḍaṅga*), of the long pepper (*pippala* [*i*], Skt. *pippalī*), of the black pepper (*marica*), of the chebulic or yellow myrobalan (*harītaka*, Skt. *harītakī*),[39] of the beleric myrobalan (*vibhītaka*, Skt. *v*[*b*]*ibhītaka*),[40] of the emblic myrobalan (*āmalaka*, Skt. *āmalakī*), and of the *goṭhaphala* (*goṭha* [var. *goṭṭha*] *phala*).[41] Any fruit that does not serve as solid or soft food can be used. The fruits may be stored indefinitely.[42]

In the medical texts, a well-developed classification of fruits (*phala*) is found under the heading *phalavarga* (class of fruits) in the chapters on foods and drinks (*annapāna*).[43] As with leaves, fruits are also listed as one of the principal sources of the 600 evacuatives.[44] The three varieties of myrobalan—yellow, beleric, and emblic—sometimes, as in the Pāli, occur together and are commonly know as *triphalā* ("three fruits") in *āyurveda*.[45]

The classification of fruits as a separate category of medicinal foods is common to both early āyurvedic and early Buddhist monastic medicine. Like leaves, they were considered as essential components of the respective pharmacopoeias.

Gums or Resins

Sick monks required gums or resins (*jatu*) as medicines. The Buddha permitted the following medicinal gums: resins from asafetida (*hiṅgu*), *hiṅgujatu* (lit. "*hiṅgu*-resin"), and *hiṅgusipāṭikā* (? Skt. *hiṅguśivāṭika*). Buddhaghosa explains that *hiṅgu*, *hiṅgujatu*, and *hiṅgusipāṭikā* are merely three types of *hiṅgu*.[46] Also permitted are *taka* (var. *takka*) (resin), *takapattī* (resin leaf?), and *takapaṇṇī* (resin leaf).[47] Buddhaghosa states that these are three kinds of lac or resin (*lākhā*, Skt. *lākṣā*).[48] Again, other resins (*sajjulasa*, Skt. *sarjarasa*)[49] may be used if they are not solid or soft food; they may be stored indefinitely.[50]

Elsewhere in the Vinaya, *jatu* occurs in the compound *jatumāsaka*, a beanlike lozenge (*māsaka*) made of gum (*jatu*).[51] Buddhaghosa states that the compound word refers to a coin made of lac (*lākhā*) or resin (*niyyāsa*) with a particular form (*rūpa*) embossed on it.[52] Likewise, the compound *jatumaṭṭhaka* (application of gum) is found in the *Bhikkhunīvibhaṅga* and is proscribed for nuns unless they are ill.[53] Gums and resins, like extracts, appear to be processed drugs rather than foods, but the Buddhists clearly understood them to be in the category of foods.

The Pāli and Sanskrit word *jatu* is used rarely in the early medical treatises. It seems to be found only in Suśruta, where the commentator Ḍalhaṇa glosses it with *lākṣā* (lac, resin).[54] In his chapter on the enumeration of drugs (*dravyasaṃgrahaṇīya*), Suśruta lists the plants of the *lākṣā* group, describes them as having an astringent, a bitter, or a sweet taste (*rasa*), and specifies the illnesses that they abate.[55] None of the identifiable resins mentioned in the Pāli is included in this list.

There is definite evidence that a group of drugs under the category of resins (*lākṣā*) developed in the tradition of Suśruta and apparently not in the other extant medical traditions. The same name for a different group of medicines occurs in the Pāli under the appellation *jatu*, a synonym of *lākṣā*. Resins as a separate category of drugs are common both to Buddhist monastic medicine and to the early medical tradition of Suśruta. In the *Suśruta Saṃhitā*, however, the names of the plants from which resin is derived differ from those found in the Pāli. This may reflect regional variation. Moreover, Suśruta's classification of resinous drugs includes a further description of their tastes and a list of the diseases they help to cure. A congruence in the classification of resins therefore occurs between the Buddhist monastic medical tradition and the *Suśruta Saṃhitā*.

Salts

Finally, sick monks needed salts (*loṇa*, Skt. *lavaṇa*) as medicines. The Buddha permitted the following five kinds of medicinal salts: ocean salt (*sāmudda*, Skt. *sāmudra*), which, Buddhaghosa says, "settles on the shore like sand"; black salt (*kāḷaloṇa*, Skt. *kālalavaṇa*), which Buddhaghosa explains as "common" salt; rock salt (*sindhava*, Skt. *saindhava*, lit. "belonging to Sindh"), which Buddhaghosa describes as "white in color"; culinary salt (*ubbhida*, Skt. *audbhida*), described by Buddhaghosa as "a sprout issuing from the earth"; and red salt (*bila*, Skt. *viḍa*), which, Buddhaghosa states, is "gathered together with all kinds of ingredients and has a red color."[56] Whatever salts are neither solid nor soft food are also permitted and can be stored indefinitely.[57]

A close parallel is found in the early medical texts. Caraka enumerates a fivefold list of salts (*lavaṇa*), very similar to that which occurs in the Pāli: *saindhava*, *sauvarcala*, *viḍa*, *audbhida*, and *sāmudra*. He explains, as a group, their properties, the medicinal preparations in which they are used, and the diseases they cure.[58] In an entirely different chapter concerning foods and drinks, Caraka enumerates the properties and healing efficacy of each of the five individually and adds to the list a sixth salt, *kālalavaṇa*.[59] Bhela mentions a group of five salts but does not list them.[60] Suśruta, in his chapter on foods and drinks, adds *romaka* to this group of five and describes, as a group, their properties and the diseases for which they are used. He then treats them individually in the same manner and adds to the basic salts *guṭikā* salts, *kuṭa* salts, and *kṣāra* salts.[61]

Both the Pāli and the early medical traditions notice five basic forms of salt (Suśruta adds one more to this list). The enumerations are essentially identical, with one exception: *kāḷaloṇa* is replaced by *sauvarcala* in the medical texts. The Sanskrit form, *kālalavaṇa*, does, however, occur in Caraka[62] and Suśruta but not in their lists of principal salts. Both Caraka and Suśruta explain that it has the "properties of *sauvarcala* without the smell."[63] The eleventh-century C.E. commentator of the *Caraka Saṃhitā*, Cakrapāṇidatta, mistakenly glosses *kālalavaṇa* with *viḍalavaṇa*.[64] The commentator of Suśruta, Ḍalhaṇa, though, states that *kālalavaṇa* is a kind of odorless *sauvarcala* salt.[65]

The salts *kālalavaṇa* and *sauvarcala* are therefore practically synonymous. This explains the appearance of *kāḷaloṇa* in the Pāli list where, according to the Sanskrit medical treatises, *sauvarcala* would be expected. The five basic salts of early *āyurveda* are therefore essentially the same as the five salts enumerated in early Buddhist medicine.

Of all the foregoing classifications of medicinal foods, that of salts best illustrates a congruence between the two traditions of Indian materia medica. Moreover, a classification of five salts as medicinal foods occurs in both the Buddhist monastic and the early āyurvedic medical tradition. In addition to the mere list, the latter provides an accumulated theoretical and practical knowledge about salts. In one section, the *Caraka Saṃhitā* gives descriptions of their properties, medical preparations in which they are used, and diseases they help to alleviate, and includes in a separate section on foods and drinks a sixth salt with a brief discussion of all six individually. The *Suśruta Saṃhitā*, in a chapter on foods and drinks, organizes its similar treatment of salts in a unified section where the five salts plus a sixth are discussed as a group and individually and the entire list is expanded to include additional salts not found in Caraka.

Although exact correspondences between Buddhist and āyurvedic materia medica are not always found, the fundamental tendency to classify and categorize medicines into distinct types is common to both. Similarities most often occur in chapters on foods and drinks (*annapāna*) in the medical treatises (especially in Suśruta), which suggests an original understanding of the substances as forms of nourishment and a secondary conception of them as medicines and drugs. This is supported by the early Buddhist medical tradition, which considered all the basic medicines as forms of food, and suggests that the earliest classification of drugs was based on ancient Indian culinary traditions. Exclusive to Buddhist monastic medicine is the list of five basic medicinal foods. Similarly, clear schemes for codifying extracts and leaves based on vegetal sources find no parallels in the medical treatises, which, however, recognizes them as important pharmaceuticals. Roots, fruits, and salts are classified according to similar schemes in both traditions, the latter two included as foods by the medical authors. The classification of a separate group of gums or resins survives only in the *Suśruta Saṃhitā.* Finally, the variation in the methods of codifying animal fat, originally a medicinal food, reflects later Hindu systematization.

The Buddhist delineation of certain foods as medicines marks an early phase in the historical evolution of Indian materia medica. A similar classification of the medicinal foods found in the early medical compendia suggests a common pharmaceutical tradition. Variations in the classificatory schemes and in the contents of each group of medicines could result from several factors, including significant influences from regional medical lore and different epistemological and religious orientations. Comprehensive discussions of drugs and their properties found in the medical treatises indicate further developments by medical theoreticians. As pharmacopoeias, called *nighaṇṭu*s, are still being produced in India, the final chapter in the evolution of Indian materia medica is yet to be written.

6

Stories of Treatments Based on Cases of Diseases

The medical section of the monastic code contains a series of stories recounting cases of sick monks and the permissible treatments that follow the rules pertaining to materia medica. These case histories provide a clear picture of medical practice current in the Buddhist monastery in the centuries preceding the common era. The ailments treated are generally minor and represent typical afflictions suffered by Buddhist cenobites in the early *saṅgha.*

This therpeutic part of Buddhist monastic medicine afforded monks and nuns a case-by-case instruction for the care of the sick. Professional medical practitioners also rendered medical aid to the ill in the *saṅgha*, indicating the close connection between medicine and Buddhist monasticism. Ākāsagotta of Rājagaha and Jīvaka Komārabhacca, whose cures will be examined in Appendix I, are two healers familiar to the Buddhists. Comparisons between Buddhist therapeutics and corresponding treatments in the early medical treatises reveal a general continuity of medical doctrines and point to a common source of medical lore. Analysis suggests that the closest connections are to the medical material compiled in the healing tradition that gave rise to the *Suśruta Saṃhitā*, but correspondences also occur with the compendia of Caraka and Bhela. Variations, however, are occasionally encountered, reflecting both the diversity of the rich storehouse of medical knowledge from which both traditions derived their medical doctines and a later Hindu veneer applied to certain therapies to render them appropriate to brāhmaṇic orthodoxy.

Buddhist monastic medicine represents the earliest extant codification of medical doctrines. Drawing on the same sources, the compilers of later medical compendia incorporated a greater quantity of medical lore, systematized it according to an etiology based on the three "peccant" humors (*doṣa*s), and classified it into eight parts (*aṅga*s) according to the different types of medicine (i.e., general medicine; major and minor surgery; toxicology; demonology; pediatrics, including obstretics; science of aphrodisiacs; and science of elixirs). A theoretical basis to medicine is merely implied in Buddhist monastic medicine, especially in the cases of the various wind diseases suffered by the monk Pilindavaccha, and of a certain monk's humoral disorder. This reflects the presence or at least the beginning of a humoral etiology and strongly indicates the Buddhist penchant for practical medical application clearly present in the following eighteen case histories.

Large Sores

A certain monk, Belaṭṭahasīsa, the preceptor of Ānanda, suffered from large sores (*thullakacchā*). The discharge (*lasikā*) from his sores caused his robes to stick to his body. The sores and scabs were moistened with water in order to loosen the robes. Buddhaghosa glosses the sores as "large eruptions" (*mahāpiḷakā*).[1]

The Buddha, after taking this matter into consideration, sanctioned the following treatments: medicinal powder (*cuṇṇa*, Skt. *cūrṇa*) was allowed for itches (*kaṇḍū̆*), which Buddhaghosa understands to be "scabs" (*kacchu*); for small eruptions (*piḷakā*, Skt. *piḍakā*), which Buddhaghosa glosses as "small bloody raised lumps, i.e. small boils"; for running sores (*assāva*, Skt. *āsrāva*), which, Buddhaghosa explains, are "a kind of fistula or urinary disorder caused by impure exudations";[2] for large sores (*thullakacchā*, var. *thullakacchu*); and for a bad-smelling body (*duggandha kāya*). For those monks suffering from less serious afflictions, however, animal dung (*chakana*, Skt. *chagana*), which Buddhaghosa glosses as "cow dung," clay (*mattikā*, Skt. *mṛttikā*), and decocted dye (*rajananipakka*) were allowed.[3] Buddhaghosa states that the last treatment involved the preparation and use of moistened, ground natural powder, perhaps as a type of soap, during a bath.[4] Finally, in order to prepare the medicines, a mortar (*udukkhala*, Skt. *ulūkhala*) and pestle (*musala*), along with a powder sifter (*cuṇṇacālānī*) and a cloth siever (*dussacālānī*), were permitted.[5] Elsewhere in the Vinaya, a type of bandage (*kaṇḍupaṭicchādi* [itch covering]), which covered the sores occurring below the navel and above the knees, was allowed for the first four afflictions treated with medicinal powder.[6]

From the Pāli accounts, *thullakacchā* appears to be a type of skin affliction, characterized by large sores or eruptions that discharge and produce scabs. It is probably a type of minor disease that the authors of the medical treatises call *kakṣā* (*kakṣyā*), a cutaneous affliction caused by bile and characterized by numerous small, large (*sthūla*), or medium-size black eruptions (*piṭakā*) resembling parched grains and located on the arms, the sides of the torso, the posterior, and the concealed parts of body (i.e., armpits and private parts).[7] *Kakṣā* is commonly identified as herpes. The Pāli term *thullakacchā* could therefore refer to the large (*sthūla*) patches of sores (*kakṣā*), that is, Sanskrit **sthūlakakṣā*, or a type of herpes characterized by large (dark) patches of sores.

Its treatment included measures prescribed for erysipelas (*visarpa*) and various skin disorders (*kuṣṭha*), specifically, applications of clarified butter (*ghṛta*) cooked with sweet drugs (*madhurauṣadha*).[8] For certain minor skin conditions (*kuṣṭha*), Caraka prescribes a medicinal powder (*cūrṇa*) combined with other medicines.[9]

Remedies for other cutaneous disorders mentioned in the Pāli—itches, small eruptions, running sores, and a bad-smelling body—find correspondences in the early medical treatises. Medicinal powder (*cūrṇa*), together with other medicines, is effective for various skin diseases (*kuṣṭha*) characterized, among other things, by itching (*kaṇḍū*) and small eruptions (*piḍakā*).[10] A poultice (*pradeha*), made of several plants, is recommended for a bad-smelling body (*śarīradaurgandhi*).[11] The types of discharges (*āsrāva*) and treatments for them are discussed at length in the *Suśruta Saṃhitā* and include discharges from skin eruptions and from the eyes.[12] The term *āsrāva* also occurs in the *Atharvaveda*, where it refers to an exudation from a wound and may also signify excessive menstrual flow or discharge.[13]

The use of animal dung, clay, and decocted dye, mentioned in the Pāli, suggests a treatment for minor skin diseases. The first two may have been employed in a plaster or a poultice, while the dye, as Buddhaghosa suggests, seems to have been a type of wash. The dung could also have been used as a medicinal scraper in the removal of scabs, and the dye employed to color the white spots that occur in certain skin diseases. A similar technique, applied in the case of cutaneous white patches, is found in the medicine of the early Vedic period.[14] In the classical āyurvedic treatises, when any of these medicines is listed in treatments for either major or minor cutaneous diseases, it is always as an ingredient along with other drugs.[15]

The similarities between the early Buddhist monastic and the early āyurvedic medical traditions concerning the treatment of skin conditions strongly suggest continuities in medical doctrine and practice. Both probably derive from a common source of medical lore.

Nonhuman Disease

A certain monk suffered from a nonhuman disease (*amanussikābādha*). He was treated but did not recover. Finally, he ate the raw flesh (*āmakamaṃsa*) and drank the raw blood (*āmakalohita*) of a swine (*sūkara*). His disease subsided. Because of this, the raw flesh and raw blood of a swine were allowed as medicines in the case of nonhuman disease.[16] Buddhaghosa understands that a nonhuman entity ate the swine's raw flesh and drank its raw blood; thereby, the disease became calm and appeased, implying that the monk was possessed by some demonic force.[17]

Chapter 60 in Suśruta's *Uttaratantra* is devoted exclusively to the warding off of nonhuman visitations (*amānuṣopasargapratiṣedha*). The group of demonic entities that attack humans is called "seizers" or "possessors" (*graha*). The Rākṣases, one of its members, are noted for their fondness for flesh and blood.[18] The nonhuman affliction of the Buddhist monk, therefore, is probably a form of possession by demonic entities, perhaps the Rākṣases.

The treatment for one possessed by such demons involved a religious healing rite in which prayers were uttered, actions performed, and offerings, which included blood and flesh, made at the proper time to the respective seizers in order to appease them. Should the incantations (*mantra*s) be ineffective in such treatments, more empirically based measures were prescribed.[19] There is no indication that the raw flesh and blood of a swine should be consumed by the victim, but surely they must be eaten to feed and appease the possessor inside the body.

The consumption of animal products, although not prohibited in the medical treatises, was an issue that generated confliciting opnions. Francis Zimmermann addresses the problem and claims that such therapeutic measures were

> inserted into a much wider medical and religious tradition that superimposes on it a *therapeutic system of purity* and nonviolence. At its own particular level, each of the two contrary principles is completely orthodox. So it is that a single doctrine can promote nonviolence, abstinence, and vegetarianism and at the same time in certain circumstances prescribe deceit, raw blood, and the flesh of carnivores.[20]

Therapeutics requiring consumption of animal flesh and blood are clear examples of extrabrāhmaṇic medical lore incorporated into the early medical compendia. The case in the Buddhist source is such a therapy. The Buddhists, uninfluenced by brāhmaṇic taboos, included the intake of swine's blood and flesh as a remedy for a disease that the *Suśruta Saṃhitā* states should be treated by means of a modified therapy: blood and flesh were offered rather than eaten to the accompaniment of *mantra*s and prayers. An application of a brāhmaṇic veneer is apparent in Suśruta's prescription. Nonhuman disease by all accounts is a case of demonic possession. The Buddhist remedy involved the consumption of animal products. The compilers of the *Suśruta Saṃhitā* adapted this remedy against nonhuman visitation in conformity with treatments for similar afflictions encountered in the magico-religious medicine of the early Vedic period. Should such remedies fail, the medical text recommends empirico-rational techniques. Buddhaghosa's explanation of the Buddhist's cure implies knowledge and acceptance of the medical tradition's remedy. Evidence of two traditions of medicine is therefore noticed in the case of demonic disease. The Buddhist monastic medical tradition preserves a treatment modified and adapted by the medical compilers along the lines of the brāhmaṇic intellectual tradition, which had already authorized Vedic medical lore.

Eye Disease

Eye diseases and their treatments receive much attention in the early Buddhist texts. A certain monk was afflicted with a disease of the eyes (*cakkhuroga*). His fellow monks helped him to calm himself. The Buddha, encountering the monks, inquired about their ailing brother and allowed the following as treatment, beginning with collyria (*añjana*):

1. Black collyrium (*kāḷañjana*, Skt. *kālāñjana*); Buddhaghosa: "one type of ointment, boiled with all ingredients."
2. *Rasañjana* (Skt. *rasāñjana*); Buddhaghosa: "it is made from several ingredients."[21]
3. "River ointment" (*sotāñjana*, Skt. *srotoñjana*); Buddhaghosa: "an ointment produced in rivers and streams, etc."[22]
4. Ocher (*geruka*, Skt. *gairika*); Buddhaghosa: "[yellow] ocher."
5. Soot (from lamps) (*kapalla* = *kajjala*, Skt. *kajjala*); Buddhaghosa: "it is taken from [that produced by] a lamp's flame."
6. Powdered collyrium (*añjanupapisana*, Skt. *añjana* + *upa* + root *pis*), made from the following:
    a. Sandalwood (*candana*); Buddhaghosa: "that beginning with red sandalwood";
    b. *Tagara*: probably the eastern India rosebay, used in eye diseases,

but also known as the Indian valerian, not noted for its use in the treatment of eye diseases;[23]

c. Black *anusārī* (*kāḷānusāriya*, Skt. *kālānusārī*);[24]
d. Silver fir (*tālīsa*): generally the leaves of this plant are used;
e. Nut grass (*bhaddamuttaka*, Skt. *bhadramusta*).

In addition to the medicines, the monks were allowed a collyrium box or tube (*añjanī*), not made of gold and silver, for storing the pulverized medicines; a lid for the box; a thread for binding; collyrium sticks (*añjanisalākā*), made from the same materials as the box, for applying the medicines; a case in which to store them; and a bag with a strap and a thread, in which to keep the box.[25]

The details concerning the medicines for the eyes and the apparatuses used to apply and to store these medicines point to the importance of eye care during this period and further suggest that a tradition of eye treatment was already well established.[26]

Suśruta's supplementary book, *Uttaratantra*, devotes its first nineteen chapters to the pathology and cure of seventy-six eye diseases (*netraroga*), which are generally of four types based on one of the three humors and their combination.[27] Nearly every disorder, whether it required surgery or not, usually has as one of its therapies the use of *añjana*s or collyria.

A special section in Chapter 18 of this late book specifies the kinds and uses of collyria and the tools and vessels used to apply and store them. After purifying the body, one of three proper collyria, which scrape (*lekhana*), heal (*ropaṇa*), or soothe (*prasādana*), is to be applied to an eye in which a particular humor is situated.[28] The ointments are threefold in form: small balls (*guṭikā* [*varti*]), juices (*rasa*) (Ḍalhaṇa: "*ghanarasa*" [thick juice]), and powders (*cūrṇa*), which, Ḍalhaṇa states, should be used respectively in afflictions of severe, medium, and weak strength.[29] Vessels (*bhājana*) and sticks (*śalākā*) should be made with qualities equal to that of the collyria. They may be fashioned from gold, silver, animal horns, copper, *vaid(ḍ)ūrya* stone (stone from Vidura, probably lapis lazuli), or bell metal (*kāṃsya*). The stick is to be bud shaped at both ends, with a circumference the size of a wild pea (*kalāya*), eight finger breadths in length, slender in the middle, well made with a convenient handle (Ḍalhaṇa: "easily grasped"). A stick made from branches of the cluster fig (*udumbara*) (Ḍalhaṇa: "copper"), from stones (Ḍalhaṇa: "*vaid*[*ḍ*]*ūrya* stone"), or from that of the body—that is, bone (Ḍalhaṇa: "horn, and so on")—should be suitable.[30]

Nearly all the medicines for eye disorders mentioned in the Pāli are found in the same context in the early medical treatises.[31] Black collyrium (*kālāñjana*) is named possibly only once in the early medical literature as a variant reading of *lājāñjana* (fried grain and collyrium) in Suśruta. Ḍalhaṇa cites the variant and glosses it with *sauvīrāñjana*, which is not found in the Pāli list of collyria.[32] *Sauvīrāñjana* occurs in the medical treatises[33] and, along with *rasāñjana* and *srotoñjana*, is one of the three basic collyria of early āyurvedic medicine (later, two other types, *puṣpāñjana* and *nīlāñjana*, were added, bringing the total to five).[34] It is described as black in color and derives its name from the mountains of Sauvīra, a country along the Indus River, whence the collyrium was obtained.[35] It is reasonable to assume, therefore, that *sauvīra* collyrium may have been known by another name to the Buddhists or perhaps a closely similar type of *añjana* was substituted for it.

Eye care was clearly a very old and well-established custom in ancient India. The similarities of basic ingredients and techniques used to treat eye diseases in the Pāli account and in the medical literature reflect a continuity in medical doctrine that dates from an early period and point to a common source of medical doctrines. Buddhist ophthalmology, however, does not mention surgery as a treatment of certain types of eye disorders, but in the *Suśruta Saṃhitā* eye surgery receives much attention and is one of the areas of medicine for which the Indians gained great acclaim.[36] Although probably known, eye surgery appears not to have been incorporated into Buddhist monastic medicine.

The Sivi Jātaka (no. 499) recounts how a certain skillful physician (*vejja*), Sīvaka, removed the eyes (*cakkhu*) of King Sivi to give them to a blind Brāhmaṇ as an act of great piety. Considering it inappropriate to operate on the eye with a surgical instrument (*sattha*), Sīvaka crushed various medicines (*nānābhesajjāni ghaṃsitvā*), treated a blue lotus with the medicinal powder (*bhesajjacuṇṇena nīluppale paribhāvetvā*), and gently stroked it over the right eye (*dakkhiṇam akkhim upasiṃghāpesi*). The eye rolled around (*akkhi parivatti*), and the king suffered great pain. The physician applied the powder a second time, and the eye began to be loosened from its socket. A third time, a sharper (*kharatara*) powder was applied, and by the power of the drug (*osadhabala*) the eye came out and hung at the end of a tendon (*nahārusutta*). The physician cut it off and gave it to the king, who in turn presented it to the Brāhmaṇ. The same procedure was followed for the removal of the left eye. With both eyes in his sockets, the Brāhmaṇ was given his sight.[37]

A well-established and old tradition of ophthalmology, codified in the early Buddhist monastic medicine, is discussed in elaborated from in the early medical treatises, which include a humoral etiology and surgical therapies wanting in Buddhist monastic medicine. A common source of ancient Indian eye care, therefore, contributed to both medical traditions.

Head Irritated by Heat (Head Disease)

The monk Pilindavaccha suffered from heat irritating his head (*sīsābhitāpa*). Three treatments were allowed: oil on top of the head (*muddhani telaka*), nasal therapy (*natthukamma*), and inhalation of smoke (*dhūmaṃ pātum*, lit. "to drink smoke"). Because there were difficulties in administering the oil, a nose spoon (*natthukaraṇī*), made of the same substances as the collyrium rods, excluding gold and silver, was allowed. Since the medicine went in the nose in uneven quantities, a double nose spoon (*yamakanatthukaraṇī*) was permitted. Buddhaghosa understands this type of nose spoon to be one nasal spoon beginning with an even flow from two tubes. As the patient showed no signs of recovery, the inhalation of smoke from a burning wick, anointed most likely with ghee, was allowed. The monk, however, burned his throat, so a smoke conductor (*dhūmanetta*), fashioned from the same materials as the nose spoon, was then allowed to be used. A cover (*apidhāna*) to prevent creatures from entering the conductors, a double bag (*yamakathavika*) to stop the conductors from rubbing together, and a shoulder strap and thread for binding (*aṃsabandhaka bandhanasuttaka*) were permitted to the monks.[38]

Details of diseases of the head found in the early medical treatises are remarkably similar to information provided in Buddhist monastic medicine. Head diseases are treated quite extensively by both Caraka and Suśruta. The former discusses them under the generic title *śiroroga* (disease of the head), while the latter employs the name *śirobhitāpa*,[39] which is equivalent to the Pāli *sīsābhitāpa* and which Ḍalhaṇa glosses as *śiroroga*.[40] Originally, there appear to have been five types of head disorders: three caused by each of the humors, one by a combination of the humors, and one by worms (*kṛmi*).[41] To these were added six other types: *śaṅkhaka*, *ardhāvabhedaka*, *anantavāta*, and *sūryāvarta*, described in terms of the vitiation of the "peccant" humors and found both in Caraka and in Suśruta,[42] and those caused by blood (*raktaja*) and bodily waste (*kṣayaja*), which occur only in Suśruta.[43]

According to Suśruta, one of the first treatments of head diseases is a strong purgation of the head with oil and honey. This was followed often by nasal therapy (*nasya* [*nasta*] *karman*), which in turn is usually followed by smoke therapy (*dhūma*) to prevent the formation of phlegm.[44]

The techniques for administering these two types of therapy are detailed by both Caraka and Suśruta and closely resemble the method used by the Buddhists. After the head has been made to sweat by rubbing, the patient should be placed in a proper, supine position and the physician should administer warm, medicated nasal oil (*sneha*) by means of a type of pipette (*praṇāḍī*) (or, according to Suśruta, a shell [*śuktī*]) or by cotton (*picu*) equally into both nostrils. Suśruta states that the nasal oil should be poured into and kept in a silver, gold, or red-earthen vessel or in an oyster shell while introducing it into the nostrils.[45]

The technique of inhalation of smoke required the use of smoke conductors or pipes (*dhūmanetra*), fashioned from the same materials as the enema tube (*vastinetra*), which included gold, silver, copper, iron, brass, ivory, animal horn, gems, heartwood, tin, or bamboo.[46] It has a prescribed length and circumference. When the pipe was used, a medicated pad (*varti*), first anointed with oil and then set ablaze, was to be placed in the mouth of the pipe. The patient, comfortably seated, well disposed, with a straight back and downcast gaze, and alert, should then inhale the smoke through the two nostrils, through the mouth, or through both the nostrils and the mouth.[47]

The threefold method of oiling, nasal therapy, and smoke therapy outlined in the Pāli is found fully detailed in the early medical literature, indicating a common origin for treatments of diseases of the head and continuity in medical doctrine. Although mentioned in both Caraka (Bhela) and Suśruta, the greater detail concerning the particulars of the nasal and smoke therapies and the instruments used to administer them is found in the *Suśruta Saṃhitā*. The head disorders described in the medical treatises rely almost entirely on a humoral-based nosology wanting in Buddhist monastic medicine.

Affliction of Wind

The monk Pilindavaccha suffered from the affliction of wind (*vātābādha*). On the recommendation of physicians (*vejja*, Skt. *vaidya*), oil (*tela*) was decocted, combined with a weakened intoxicating drink (*majja*, Skt. *mada*, *madya*), and given to the monk. If the drink was too strong, an oily massage (*abbhañjana*) was to be administered. Vessels made of copper, wood, or fruit were used as receptacles to hold the boiled oil.[48]

The meager amount of information offered in the Pāli allows us to posit only general statements concerning a connection with the early medical traditions. The medical treatises of Caraka and Suśruta describe numerous wind disorders (*vātavikāra*) or wind diseases (*vātaroga*, *vātavyādhi*) based on the humoral conception of the derangement of wind (*vāta*, *vāyu*), alone or combined with bile and phlegm, localized in different parts of the body.[49]

The authors of both medical treatises mention several therapies for wind diseases. Nearly every one requires a decoction of oil (*taila*) combined with other medicines, to be administered to the patient internally, externally, or in an enema. Unguents are also mentioned in certain treatments.[50] Caraka emphasizes the importance of decocted oil, stating, "There is not anything better than oil [*taila*].... After processing, it is more powerful. Therefore, cooked a hundred or a thousand times with [drugs of] the wind [*vāta*]-removing group, it very quickly destroys the toxins situated in the smallest paths."[51] The particularly efficacious oil remedy called "cooked a thousand times" (*sahasrapāka*) is to be put, after it has been prepared, in a receptacle made of gold, silver, or earth.[52]

No clearly defined remedy corresponding to that given to Pilindavaccha, however, is found in the early medical books, although the basic ingredient of oil (*tela*, *taila*) occurring in both points to a common origin. The Pāli word *vejja* implies medical practitioners (Skt. *vaidya*) closely associated with the ascetic Buddhists and their monastic community. Their recommendations for the treatment of the disease affecting the wind were based on medical doctrines utilized by the Buddhists and codified in the early medical treatises. The mention of physicians in this context helps to establish the close link between Buddhist monasticism and ancient Indian medicine, both of which derived their medical doctrines essentially from the same sources.

Wind in the Limbs

The monk Pilindavaccha suffered from wind in the limbs (*aṅgavāta*), which Buddhaghosa explains as "wind in all the limbs." The treatment employed involved a series of actions causing the patient to sweat. A sweating treatment, or sudation (*sedakamma*), was performed. The monk's condition did not improve, so the following actions were taken: a sweating by use of provisions (artificial means) (*sambhāraseda*), which Buddhaghosa explains as sweating by the use of different kinds of leaves and sprouts; and a great sweating treatment (*mahāseda*), the procedure for which Buddhaghosa explains as follows: "Having filled a pit the size of a man with burning coals, having covered it with standard dirt, having strewn the place with various kinds of leaves that remove wind, and having lain the body smeared with oil in that place, the body is sweated by rolling in it." Since the monk still had not recovered, a treatment of water with sprouts (and leaves) (*bhaṅgodaka*)[53] was administered. Buddhaghosa explains that the water was boiled with different leaves and sprouts, and describes the process as a sweating of the patient, who is repeatedly sprinkled with water and leaves. As the monk still did not improve, a final treatment by means of water storeroom (*udakakoṭṭhaka*) was given. Buddhaghosa states that this was a sweating treatment permitted after the patient enters a vessel or a tub filled with hot water; that is, it was sweating treatment resulting from soaking in hot water.[54]

The medical treatises do not have the precise equivalent to the Pāli *aṅgavāta* (wind in the limbs). Caraka lists as part of the wind diseases "wind excited in all the limbs" (*sarvāṅgakupita vāta*) and states that sweating or sudation (*sveda*, Pāli *seda*) is used in diseases (*gada*) characterized by wind (*vāta*) and phlegm (*kapha*). He goes on to explain that sudations are also used in afflictions affecting all limbs (*sarvāṅga vikāra*).[55]

Caraka enumerates thirteen types of sudations (*svedakarman*, Pāli *sedakamma*), while Suśruta divides them into four types that include most of those listed in Caraka. Bhela speaks of eight types and uses the expression *svedakarman* (sweating treatment) in very general terms, not specific to wind disorders.[56]

Definite similarities occur between the early medical tradition and that of the early Buddhists. The remarks of Buddhaghosa are especially significant in this regard. His definition of the great sudation seems to be based on Suśruta's description of a form of sweating by indirect heat (*ūṣmasveda*): "Having dug up earth measuring the length of a man, having heated it with wood form the catechu tree [*khadira*], having sprinkled it with milk, fermented rice water, and water, and having covered it with leaves and sprouts [*patrabhaṅga*],[57] one should cause him to sweat, there, while he lies down."[58] Ḍalhaṇa calls this "earth sudation" (*bhūsveda*), a term used by Caraka.[59] Bhela, however, does not include this type of treatment in his list of eight forms of sudation.[60] Likewise, Buddhaghosa's description of a treatment by water with sprouts (and leaves) closely resembles the medical treatises' "sprinkling sudation" (*pariṣekasveda*). It required the use of a multiholed container filled with a warm decoction of wind-destroying drugs, which was sprinkled or showered over the patient.[61]

Finally, both the canonical reference to and Buddhaghosa's explanation of the treatment by a water storeroom (*udakakoṭṭaka*) corresponds to the medical authors' "bath sudation" (*avagāhasveda*). This treatment involved a sweating by taking a bath in a storeroom or chamber (*koṣṭhaka*, Pāli *koṭṭhaka*) filled with wind-destroying drugs and any one of several warm liquids, which included hot water (*uṣṇasalila*).[62] Bhela, closer to the Pāli, describes a sudation by means of a "water storeroom" (*udakoṣṭha*) as follows: "One should cause to sweat him who is to be sweated, after he has entered a caldron [*kaṭāha*] half-filled with water (warmed and purified)."[63]

The early Buddhist monastic medical prescription of treatment by means of sudations has close correspondence to the techniques detailed in treatises of the early medical tradition. The Pāli name of the disease, *aṅgavāta*, is not clearly expressed in the medical treatises, nor is an equivalent to the Pāli *mahāseda* found. These variants point to obvious differences between the transmitted textual and scholastic traditions and the then-current practical knowledge of medicine in ancient India. By Buddhaghosa's time (fifth century C.E.), however, knowledge of sudations was widespread in Buddhist circles. It is likely that most of his information about this form of treatment derives from the *Suśruta Saṃhitā*, which by that time probably existed in its extant form.

Wind in the Joints

The monk Pilindavaccha suffered from a third type of wind disease, "wind in the joints" (*pabbavāta*), which Buddhaghosa defines as "wind piercing (or striking) every joint." First, blood was let from him (*lohitam mocetum*, lit. "to release blood"). According to Buddhaghosa, this was accomplished by means of a kinfe. He did not recover, so that after letting blood, cupping by means of a horn (*visāṇena gahetum*, lit. "to cause to seize by a horn") was employed.[64]

The disease of wind in the joints (*sandhi*) is included in the list of wind disorders enumerated by both Caraka and Suśruta, and is a special symptom of wind in the blood (*vātaśoṇita* or *vātarakta*).[65] Caraka explains that it is siutated in both hands, the feet, the fingers, and all the joints (*sarvasandhi*). Having made its root (*mūla*) in the hands and feet, it then spreads throughout the body.[66]

The therapeutic procedure of bloodletting (*raktamokṣaṇa*, lit. "releasing blood") is mentioned by both Caraka and Suśruta. Suśruta details the procedures for letting blood with a surgical instrument (*śastravisrāvaṇa*) and divides them into two types: scarification (*pracchāna*) and venesection (*sirāvyadhana*, lit. "piercing the blood vessel"). Likewise, he devotes an entire chapter to venesection, at the end of which he speaks of scarification by means of tubes, horns (*viṣāṇa*), gourds, and leeches (*jalauka*).[67]

For the treatment of wind in the joints, ligaments, and bones, Suśruta prescribes the use of fats, poultices, cauterizations, ligatures, and massage.

He recommends bloodletting (*asṛgvimokṣaṇa*) in cases where the wind is situated in skin, muscle, blood, or blood vessels. More generally, when wind has lodged in all limbs (*sarvāṅga*), he prescribes bloodletting (*sirāmokṣa*), along with various types of sudation. "When [the wind] is situated in one limb [*ekāṅga*]," Suśruta says, "the wise [physician] should conquer it by means of [cupping with] a horn [*śṛṅga*]." Likewise, the primary treatment of *vātarakta* is a constant and gradual bleeding (*avasiñcet*; glossed by Ḍalhaṇa as *srāvayet*), followed by an application of plasters.[68] Caraka recommends bloodletting by means of scarification or venesection with a horn, leech, needle, or bottle gourd when wind obstructs the passage of blood and enters the joints.[69]

Like the two previous diseases affecting wind (*vāta*), this one ("wind in the joints") has a close affinity with the appropriate teachings mentioned in the early medical treatises. The Pāli names *aṅgavāta* (wind in the limbs) and *pabbavāta* (wind in the joints) quite likely correspond respectively to the medical authors' "wind in all limbs (or parts)" and "wind in one limb (or part)." In terms of treatment, oil decoction was good for general wind, sweating or sudation for wind in all parts, and bloodletting for wind in a single part of the body.

The Buddhist monastic medical explanation of wind disorders in this manner suggests that a medical tradition in which these types of diseases were well known existed at or before the time the Vinaya rules were codified. It appears that this tradition was very close to that which gave rise to the medical teachings of Caraka and Suśruta. The emphasis placed on wind (*vāta*) points to the significant place it held in the early formulations of an etiology based on the "peccant" humors. Incorporation of bloodletting into Buddhist monastic medicine is significant. Because the procedure involved contact with blood, a most polluting substance, higher orders of the brāhmaṇic society would not have readily accepted it. Being part of the Buddhist and early āyurvedic medical traditions, bloodletting was probably a therapy included in the early storehouse of medical lore. Moreover, common to both ancient Indian and ancient Hellenistic and Chinese medical systems, bleeding was a medical technique transcending cultural and geographical boundaries, implying perhaps a transmission of medical knowledge.

Split-open Feet

The monk Pilindavaccha's feet became split open (*pādā phālitā*). His treatment involved both a "foot massage" (*pādabbhañjana*) and a traditional foot wash (*pajja*, Skt. *padya*).[70] Buddhaghosa reads *majja* (intoxicating drink) for *pajja* and explains that when the feet become naturally split open (i.e., opened on their own), monks are allowed to prepare an intoxicating drink, having put various medicines in coconuts and so on, and to boil the medicines beneficial to the feet.[71]

The medical treatises speak of an affliction of split-open feet in two separate ways according to the particular tradition. Suśruta describes the affliction of split-open feet (*pādadārikā* or *pādadārī*) as follows: "The wind of one accustomed to wandering, being situated in the soles of both extremely dry feet, causes painful splitting (*dārī*)." it is treated first by bleeding, followed by sweating and (foot) massage, and finally by applying a medicated foot plaster (*pādalepa*). Ḍalhaṇa states that the sweating and massage are carried out before the letting of blood.[72] This makes better sense. Caraka mentions that the affliction of the cracking of (the soles of) the feet (*pāda sphuṭana*) is prevented by a foot massage (*pādābhyaṅga*) with oil.[73] Likewise, Suśruta states that a foot massage (*pādābhyaṅga*) softens the skin (of the soles) of the feet (*pādatvaṁmṛdukārin*).[74]

Neither of these treatments corresponds exactly to the Buddhist canonical account. The use of foot massage with oil, however, is the more correct meaning of the Pāli expression *pādabbhañjana*. Rather than understanding the compound as *pādaby*(*vy*)*añjana* (foot unction),[75] it should be *pādābhyañjana*, (foot massage). In the medical tradition, *abhyañjana* is equivalent to *snehābhyaṅga* (massage with oil),[76] so that *pādābhyañjana* is the same as *pādābhyaṅga* (foot massage with oil), which is a principal means of preventing and healing cracked soles in both Caraka and Suśruta.[77]

Translators have rendered the Pāli term *pajja* as "foot salve."[78] Its literal meaning is "related to the foot," and elsewhere in the Pāli canon, the word occurs with *udaka* (water). It refers to an old custom offered a guest or a stranger who has been welcomed in the house. First a seat (*āsana*) was prepared for him, and then his feet were cared for by means of washing them with scented water and massaging them with sesame oil cooked 100 times.[79]

The treatment of the monk's split-open feet obviously was based on this long-standing custom. The word *pajja* in the passage, therefore, implies the missing *udaka* and refers to that part of the traditional method of foot care utilizing scented water as a foot wash. The other part involving a foot massage is already mentioned in the Vinaya passage in the phrase *pādabbhañjana*.

Suśruta's description of the affliction of split-open feet seems to fit especially a problem that a Buddhist monk might readily encounter.

Monks were known for their mendicant and wandering activities, thus making them particularly susceptible to afflictions of the feet. Walking barefoot can result in the growth of a thick horny layer on the soles that is prone to break open, often permitting infections to develop. Suśruta also states that the disorder is caused by wind (*vāta*). Although no mention of a humoral basis for the ailment is expressed in the Pāli, one could be inferred from its proximity to the previously mentioned wind diseases in this Vinaya documentation of early Buddhist monastic medicine. This disease is the fourth and final disorder suffered by the monk Pilindavaccha. If the compiler was thinking in terms of grouping the diseases according to humoral causes, it would be logical to place this wind-caused disease along with the others, using Pilindavaccha as the patient and the thread trying them together. A variant, however, does occur: bloodletting, a principal form of treatment for split feet in the medical texts, is not mentioned in the Pāli account but was nevertheless known to the early Buddhists.

Elsewhere in the Vinaya, a monk suffered from an affliction of eruptions on his feet (*pādakhīlābādha*). As a result of this, monks were permitted to wear sandals (*upāhana*) when their feet were painful (*dukkha*), split-open (*phālita*), or afflicted with eruptions (*pādakhīla*).[80]

The medical texts do not mention the specific afflictions of the feet for which the use of footwear is beneficial but Caraka states that wearing of foot coverings (*pādatradhāraṇa*), among other things, wards off injuries (or calamities) of the two feet (*pādayor vyasanāpaha*).[81]

The treatment of split-open feet by means of a foot massage and the use of footgear to protect the feet from injury derive from the tradition of wandering ascetics. They were codified in the Buddhist monastic medical tradition and incorporated into the early medical treatises, which included other therapies from traditions of ancient Indian medical lore.

Swellings

A certain monk had an affliction of swellings (*gaṇḍa*). The following treatments were administered: (1) lancing or treating with a knife (*satthakamma*); (2) using water with medical decoction (*kasāvodaka*); (3) applying a paste of sesame (*tilakakka*), which Buddhaghosa says is made from ground sesame seeds; (4) placing a bandage or compress (*kabaḷikā*), which Buddhaghosa describes as a lump of barely meal, in the open wound; (5) binding a rag (dressing) around the wound (*vaṇabandhanacola*); (6) sprinkling mustard powder (?) (*sāsapakuṭṭa*) on the wound (*vaṇa*) to stop it from itching (*kaṇḍu*) (Buddhaghosa glosses the powder as "ground mustard"); (7) using smoke (*dhūma*) to treat the festering wound (*vaṇo kilijjittha*, lit. "the wound became inflamed"); (8) cutting off with a piece of salt crystal (*loṇasakkharikāya chinditum*) when the flesh of the wound stood up (*vaṇamaṃsaṃ vuṭṭhāti*), which Buddhaghosa explains as "the additional flesh [covering] arises like a peg"; (9) applying sesame oil (*vaṇatela*, lit. "wound oil") on a wound that does not heal (*vaṇo na rūhati*); (10) using a linen bandage (*vikāsika*) to stop the running oil (*telaṃ galati*), which Buddhaghosa says is "a small piece of cloth that restrains the oil"; and, finally, (11) trying every cure for a wound (*sabbaṃ vaṇapaṭikamma*), which Buddhaghosa defines as "whatever cure for a wound there is."[82]

This treatment of swelling (*gaṇḍa*) is significant because it provides a detailed course of action that, in respect to the individual elements, has close parallels in the medical texts, but with respect to the sequence of steps, exact correspondences are wanting.

The word *gaṇḍa* maintains a significant place in Pāli terminology pertaining to human disease and injury and in similes referring to bodily states. Elsewhere in the Vinaya, it is found in a list of five diseases prevalent among the people of Magadha: skin disease (*kuṭṭha*, Skt. *kuṣṭha*), swellings (*gaṇḍa*), cutaneous white patches (*kilāsa*), consumption (*sosa*, Skt. *śoṣa*, lit. "drying up"), and epilepsy (*apamāra*, Skt. *apasmāra*). Monks afflicted with any of these ailments were not allowed to go forth for alms.[83] Occasionally, diabetes (*madhumehika*, lit. "honey urine") is also mentioned with the five.[84] It was considered an offense of expiation if nuns allowed swellings (*gaṇḍa*) or scabs (*rūhita*) on the lower part of the body to burst or to be treated by another.[85]

Swellings of various kinds were very common and, because of their morbid qualities, were often symbolically used in Buddhist similes that elucidate the different types of hindrances to proper religious growth.[86]

The Pāli and Sanskrit term *gaṇḍa* appears only rarely in the medical treatises. It is defined as an affliction of the flesh (*māṃsa pradoṣaja*) and is mentioned along with other skin disorders, which are enumerated and discussed in detail by Suśruta.[87]

The general treatment prescribed for skin disorders follows closely that recommended for *vraṇa* (wounds, or sores), a word that is frequently encountered in the above description of treatment by the Buddhists and is the same as the Pāli *vaṇa*. Both Caraka and Suśruta have very similar chapters devoted to the various methods to cure wounds.[88] An examination of these chapters shows that almost all the remedies and procedures are presented in the Buddhist account but they are arranged differently. Caraka enumerates thirty-six types of treatments for wounds, including six surgical operations (*śastrakarman*), among which are found scraping, cleansing, and healing decoctions (*śodhanaropaṇyau kaṣāyau*); two types of oil (*dve taile*); two types of bandages (*dve bhandhane*), which include one made of linen (*kṣauma*); hardening and softening fumigations and plasters (*kāṭhinyamārdavakare dhūpanālepane*); and powdering the wound (*vraṇāvacūrṇaṇa*). Also found among these fundamental actions are two types of cauterization (*dāho dvividhaḥ*)—that is with fire and with alkali.[89]

Suśruta's recommended treatment is very similar but has sixty rather than thirty-six specific therapeutic actions.[90] The following significant variants occur in Suśruta and bear a close connection to the Buddhist's treatment: scraping off of wounds with rock salt is prescribed in certain instances; sesame oil combined with mustard oil (*sarṣapasnehayuktena*... *tailena*) is used for raised, dry wounds with slight secretion; and a cleansing paste (*śodhana kalka*), made of sesame and other ingredients, is prescribed to heal the wound. Here *kalka* is equivalent to Pāli *kakka* in the compound *tilakakka* (sesame-oil paste). Both Caraka and Suśruta recommend fumigation (*dhūpana*). Suśruta specifies that it be used for wind-caused wounds, accompanied by heavy discharge. Like Caraka, Suśruta prescribes the use of cauterization with fire or alkali.[91] Nowhere in the Pāli account is the use of cauterization mentioned. Similarly, both Caraka and Suśruta state that one of the first courses of action in the treatment of wounds and similar afflictions is to let blood.[92] No method of bleeding is prescribed in the Pāli. Although these therapies are not stated in the Buddhist account, they may be implied by the final course of action—that is, the use of every cure for a wound.

From the examination of the treatments for wounds in the early Sanskrit medical treatises, we notice that nearly all the techniques given in the Pāli are found in the medical tradition, with the closer affinity being to Suśruta than to Caraka. Certain specific measures, such as the appliction of sesame oil for a wound that does not close and the use of a linen bandage to stop the running oil, appear to be wanting in the medical texts. In the two chapters on the treatment of wounds in Caraka and Suśruta, sesame oil, either in a paste or in a decoction, seems to have been prescribed to help promote the closing of the wound. Likewise, an exact parallel to the course of treatment outlined in the Buddhist canonical account is not found in the early medical tradition, suggesting that the Buddhist codification of the treatment involved a compilation of therapeutic measures that evidently derived from the common storehouse of medical lore utilized by both the early medical authors and the early Buddhists in their respective formulations of medical doctrines pertaining to wounds and swellings. The Buddhist use of the word *gaṇḍa* (swelling) to name the affliction, followed by the constant use of the word *vaṇa* (wound) in the course of the treatments, points to a connection between the two external maladies and implies that they were considered as synonyms. The medical authors' more detailed classification of wounds and swellings reflects a later development and a deeper technical understanding, which may well have been added during the several redactions of the texts.

Snakebite

A certain monk suffered from snakebite (*ahinā daṭṭha*, lit. "bitten by a snake"). The treatment involved the use of the four great foul things (*cattāri mahāvikaṭāni*), which, elsewhere in the Vinaya, are used for cleaning teeth and are not considered to be substantial nutrition:[93] (1) dung (*gūtha*), (2) urine (*mutta*), (3) ashes (*chārika*), and (4) clay or dirt (*mattikā*).[94]

In the Vinaya, two other cases of snakebite (*ahinā daṭṭha*) are encountered. The first, suffered by a certain monk, required fire (*aggi*) to be used in the treatment.[95]

In the second case, a certain monk was bitten by a snake and died. The Buddha was consulted and explained that the monk died because he did not have kind thoughts (*mettā cittā*) toward the four royal families of snakes (*ahirājakula*): Virūpakkha, Erāpatha, Chabyāputta, and Kaṇhāgotamaka. He therefore allowed kind thoughts to be directed toward the snakes and the following protective charm (*parittā*) to be recited:

> Kindness from me to the Virūpakkhas; kindness from me to the Erāpathas; kindness from me to the Chabyāputtas; kindness from me to the Kaṇhāgotamakas; kindness from me to the footless ones; kindness from me to the two-footed ones; kindness from me to the four-footed ones; kindness from me to the many-footed ones.
>
> Let not the footless one harm me; let not the two-footed one harm me; let not the four-footed one harm me; let not the many-footed one harm me. Let all sentient beings, all breathing beings, and all living beings, totally, see all auspicious things; let not any evil come [i.e., attack].
>
> Limitless is the Buddha, limitless is the Dhamma, limitless is the Saṅgha. Limited are the creatures that creep: snakes, scorpions, centipedes, spiders, lizards, and mice.
>
> I have made protection; I have made a charm. Let the living beings retreat. Surely, I make obeisance to the Lord [Buddha]; I make obeisance to the seven fully awakened ones.

In addition to the charm, the Buddha permitted the more empirical and rational therapy of bleeding (*lohitam mocetum*).[96]

The medical authors devote entire chapters to the identification and classification of various venomous snakes, types of bite, and their symptoms and treatments.[97] Caraka lists twenty-five basic treatments for snakebite, and Suśruta's course of treatment follows the same general lines as Caraka's.[98] The general remedies incorporate, like Buddhist monastic medicine, both magico-religious and empirico-rational medicine. The former involved the recitation of incantations (*mantra*s) and the performance of religious rites, while the latter required the use of tourniquets, bloodletting, cauterization, sucking, scarification, plasters of antidotes (*agada*), purgatives, emetics, and nasal therapy.

The use of fire in one of the cases of snakebite found in the Vinaya probably implies cauterization but could also refer to the ritual use of fire. In the latter, the snake might have been coerced by fire and charms into sucking back its venom. Such a magico-religious use of fire against snakebite is recorded elsewhere in the Pāli canon.[99]

The enumeration of the four royal families of snakes, the recitation of an incantation, and the use of bloodletting, found in the third case, correspond closely to the general course of treatment prescribed in the medical texts. In the first example of snakebite, the four foul substances appear to have served as food for those non-Buddhist rivals, such as the Ājīvikas, who engaged in extreme asceticism.[100] One or more of the substances is occasionally encountered in treatments against poisons in the medical texts. The four are, however, never found together under the appellation "the four great foul things." The significance that these substances held for the Buddhists in the treatment of snakebite is not paralleled in the early medical tradition. The connection with the śramaṇic traditions through the Ājīvikas, however, points to medicines and treatments of the ascetic physicians. Moreover, putrid urine (of cattle) as medicine (*pūtimuttabhesajja*), as previously mentioned, was one of the Buddhists' four principal resources of life. Originating with the ascetic traditions, the use of cow's urine and dung as medicines was included in early Buddhist monastic medicine.[101]

A series of verses addressing the treatment of snakebite occurring near the end of Caraka's chapter on poison states that if one is bitten, "one should immediately bite that snake or else a clod of earth (*loṣṭa*)" and then apply a tourniquet and use excision or cauterization. As a further part of the treatment, different stones and vegetal amulets should be worn and auspicious birds kept.[102]

Suśruta mentions that in the absence of the appropriate antidotes to be taken internally, a solution of black clay form a termite's nest (*kṛṣṇa valmīkamṛttikā*) may be used.[103] A general antidote called *kṣārāgada* (caustic antidote) included the ashes (*bhasman*) from various plants dissolved in cow's urine (*gavām mūtra*) and boiled with numerous other powdered drugs and ingredients.[104] The urine of various animals was considered to be purifying and was used for a number of ailments, including poisoning (*viṣa*).[105] The juices of cow dung (*gomayarasa*) and ash (*bhasman*) or earth (*mṛda*) are mentioned in a particular treatment against poisoning.[106]

None of the early medical treatises specifies that dung, urine, ashes, and clay or earth should be used together as the principal ingredients in any cure for snakebite. Likewise, the terminology used in the Pāli differs somewhat from the Sanskrit in the words for "snake," for "dung," for "ashes," and perhaps also for "clay" or "dirt." The word *ahi* (snake) is of Vedic origin, being replaced generally by *sarpa* in classical Sanskrit. The tradition that gave rise to the Buddhist cure for snakebite appears to be slightly different from that which is expressed in the early medical texts, yet aspects of magico-religious healing are common to both. The ingredients that make up the Buddhist cure probably derive from an older śramaṇic medical tradition, of which the Pāli formula is the earliest codification. By the time the medical treatises reached their final redaction, certain toxicological traditions were included, while others were omitted.

The references to snakebite treated by incantation and bloodletting, however, reflects a closer connection with an āyurvedic medical tradition. Resorting to empirico-rational therapeutics when magico-religious techniques fail occurs in both the Buddhist monastic and the āyurvedic medical traditions.[107] The toxicological tradition is very ancient in India. Clear distinctions, however, can be ascertained between the use of magic (ritual actions and incantations), which, based on similarities in the *Atharvaveda*, is perhaps the older, and the use of more empirical methods (bloodletting, etc.), which, because the details of such practices first appear in the early āyurvedic texts, are probably more recent. Both approaches have been preserved in the early medical works and in the Pāli canon and suggest a common origin.

The Effects of Harmful Drink

Two cases are reported in which monks are said to have suffered from the effects of harmful drink. In the first, a certain monk drank poison (*visaṃ pīta*). The treatment consisted of drinking (a decoction made of) dung (*gūthaṃ pāyetum*).[108] In the second, a certain monk suffered from affliction resulting from being given fabricated (artificial) poison (*gharadinnakābādha*), which Buddhaghosa states is "a disease having occurred from a drink that makes one subjected to another," that is, the result of sorcery or witchcraft. The treatment for this affliction required the drinking of (a decoction of) mud from a furrow (*sītāloḷiṃ pāyetum*),[109] which, according to Buddhaghosa, is a mixture of water and mud that sticks to a plowshare used for tilling.

The medical treatises pay particular attention to the ingestion of poison, as the king was a special target for such a mode of assassination. Detailed consideration is given to the methods by which poisons can be consumed, the symptoms of each type, and the ways of detecting poisoned food and drink.[110] Suśruta even devotes a section to proper procedures to be followed in the royal kitchen to prevent poisoning. The treament for one who has ingested poison is also included in his discussions.[111] In no instance is a treatment involving the drinking of a decoction of dung prescribed. As we noticed in the previous case of poisoning by snakebite, the juices of cow dung were employed in a certain remedy. Dung, however, does not seem to have had the significance in early *āyurveda* that it had in early Buddhist medicine. This suggests a slightly different tradition of toxicology, only part of which was incorporated into the treatises of the early medical tradition.

The second type of poisoning resulted from being given artificial poison (*ghara*).[112] Caraka classifies poisons into three basic types based on their origins: animal, plant, and *gara*.[113] Sanskrit *gara* (Pāli *ghara*) is understood to be an artificial or a fabricated poison (*kṛtrima*), made from a combination of poisonous and nonpoisonous substances.[114] Because its fatal action is prolonged, it was considered to be particularly effective against enemies, who were poisoned by putting the *gara* poison in their food. The most effective remedy for one poisoned by this means involved the use of powdered copper (*tāmrarajas*) together with honey as an emetic, and the giving of powdered gold (i.e., gold dust) (*hemacūrṇa*).[115]

The Buddhist treatment of a decoction of mud from a furrow is not found in the early medical texts. As Buddhists were not allowed to have gold in any form, the use of gold dust in medical treatment would not have been permitted. Copper, however, was not forbidden but may not have been readily available. Both gold and copper are products of the earth, suggesting perhaps that the mud from a furrow was an appropriate substitute.

In the treatments of poisons encountered in the cases of snakebite and the effects of harmful drink, the Buddhist therapies in all but one instance are sufficiently different from those of early āyurvedic medicine, as represented in the treatises of Caraka and Suśruta, to suggest that another or several traditions of toxicology were current in India around the time of the compilation of the Vinaya rules. Some of the treatments in a certain form found their way into the medical texts, while others, perhaps unknown to the compilers, were omitted or considered vile and inappropriate for a medical tradition under the domain of the Brāhmaṇs and systematically eliminated from the compendia before they attained their extant forms. Evidence of some of these other treatments seems to have survived in early Buddhist medicine. The other traditions may well have been those that were practiced and maintained in certain śramaṇic groups, including, among others, the Ājīvikas.

Defective Digestion

A certain monk suffered from defective digestion (*duṭṭhagahaṇika*), which Buddhaghosa understands to be "failing digestion; that is, feces issue forth with difficulty." The treatment involved the drinking of a solution of raw alkali (*āmisakkāraṃ pāyetum*). Buddhaghosa explains the procedure for preparing the alkali: "after having burned dry gruel, an alkaline solution flows from the ash."[116]

The early medical texts have individual chapters or sections devoted to *grahaṇī*,[117] which, corresponding to the Pāli *gahaṇika*, is understood to be an internal organ, the seat of the digestive fire, in which a disease by the same name finds its origin:

> *grahaṇī* [the seizer] is understood to be the abode of the digestive fire [*agni*] because it seizes [undigested] food. It is [located] above the navel and is supported and furthered by the power of the digestive fire. It checks undigested food and releases digested [food] from the side; but when the digestive fire is weak, it becomes defective [*duṣṭa*, Pāli *duṭṭha*] [and] emits undigested [food].[118]

The disease, precisely as Buddhaghosa claims, affects digestion and produces a type of chronic diarrhea.[119] This affliction, brought on by activities in which Buddhist monks were known to have engaged, such as restricting the diet, abstaining from food, and excessive traveling, was classified into four types based on the Indian humoral theory (i.e., one type for each of the three humors and one for a combination of them).[120]

The principal means of treatment required evacuation, followed by the administration of remedies to improve the digestive fire.[121] One of the main ways to tone up the digestive system involved the giving of alkaline mixtures (or solutions) in the form of a drink. The preparation and administration of these are detailed in Caraka and to a lesser extent in Bhela.[122] The alkali was obtained by burning leaves and other substances, a method similar to that prescribed by Buddhaghosa.

Suśruta, curiously, has not given as much attention to *grahaṇī* as has Caraka (and Bhela). He treats the disease in his chapter on diarrhea (*atisāra*). His explanations follow those of Caraka. His discussion of treatments, however, is very abbreviated, running to only five verses.[123] In them he states that the patient should be evacuated and then should be given drugs, used with various liquids, to promote appetite and digestion. Remedies mentioned in connection with the treatment of worms (in the bowels), internal (abdominal) tumors (*gulma*), ascites, and hemorrhoids, some of which use alkaline substances,[124] are to be employed in this connection. Although all the salient points concering the disease *grahaṇī* are covered by Suśruta, the brevity with which they are treated suggests that a detailed understanding of the disease and its curve derives from the early tradition of Caraka (and Bhela) rather than from that of Suśruta.

The Buddhist monastic medical tradition, as expressed both by the canonical account and by its commentary, definitely corresponds to the early āyurvedic understanding of the disease *grahaṇī* and reflects a continuity in medical doctrine. The importance given to a treatment by means of drinking alkaline solution, however, points to Caraka rather than to Suśruta. A common storehouse of medical knowledge pertaining to this disease and its treatment was the source for both the Buddhist monastic and the āyurvedic medical traditions.

Morbid Pallor or Jaundice

A certain monk suffered from morbid pallor or jaundice (*paṇḍuroga*), for which he was given a solution of urine and yellow myrobalan to drink (*muttaharītakaṃ pāyetum*).[125] Buddhaghosa explains that the treatment consisted of "yellow myrobalan mixed with cow's urine."

In the early medical treatises, an entire chapter is devoted to the symptoms and treatment of morbid pallor (*pāṇḍuroga*), of which there are either four, five, or eight types classified principally according to the humoral theory.[126]

Both Caraka and Suśruta agree that the first and most important treatment for *pāṇḍuroga* is the administration of a drug that evacuates the patient. Two principal remedies for accomplishing this are outlined: those using clarified butter (*ghṛta*, *sarpis*) as a base, and those using the urine (*mūtra*) of cows (*go*) or buffaloes (*māhiṣa*) as a base.[127]

Suśruta describes the remedies requiring the use of urine (*mūtra*): a quantity of turpeth tree (*nikumbha*), cooked in urine (Ḍalhaṇa: "of a female buffalo"), is to be drunk; cow's urine, combined with the powder of the three myrobalans (*triphalā*) and powdered iron, is to be licked (Ḍalhaṇa) many times;[128] various drugs and minerals, including yellow myrobalan (*pathyā*) decocted with cow's urine, are given to cure morbid pallor; and bowstring hemp (*mūrvā*), turmeric (*haridrā*), and emblic myrobalan (*āmalaka*), after being soaked in cow's urine for seven days, should be licked.[129] Caraka's treatments with urine are similar but more numerous, totaling eight.[130] Two of these, wanting in Suśruta, specifically prescribe that yellow myrobalan (*harītakī*) and cow's urine (*gomūtra*) should be drunk.[131]

The means of treating morbid pallor mentioned in Buddhist monastic medicine has correspondence in the two early medical treatises, illustrating a continuity in medical doctrine and pointing to a common source for the medical traditions of Caraka and Suśruta, as well as that of the early Buddhists. It is curious, although perhaps merely coincidental, that both the preceding illness, *gahaṇika*, and this one, *paṇḍuroga*, follow in the same order in the chapter on medicines (*Bhesajjakkhandha*) of the *Mahāvagga* as in the book of therapeutics (*Cikitsāsthāna*) of the *Caraka Saṃhitā*. Both are also found in an appended chapter (*Uttaratantra*) of the *Suśruta Saṃhitā* but do not follow one after the other.

Corruption of the Skin

A certain monk suffered from a corruption of the skin (*chavidosa*), which was treated by applying a scented paste (*gandhālepaṃ kātum*).[132] The expression *chavidosa* is not found in the early medical treatises. Its equivalent, *tvagdoṣa* (corruption of the skin), does occur and represents all skin disorders, including the broad category of cutaneous afflictions known as *kuṣṭha*.[133] Caraka, Bhela, and Suśruta devote two chapters each to the etiology and treatment of skin diseases (*kuṣṭha*), which are understood to be corruptions of the skin caused by the three "peccant" humors (*doṣa*s).[134] There are eighteen varieties divided into four major (*mahant*) and eleven minor (*kṣudra*) types.[135]

The first part of the treatment of *kuṣṭha* required the patient to be evacuated, followed by bloodletting when needed.[136] This was generally followed by the application of medicated pastes (*pralepa*, *lepa*, *ālepa*, or *ālepana*). The prescriptions of their principal kinds are detailed both by Caraka and by Suśruta.[137] Among these and others mentioned by the medical authors, a specifically scented type is wanting. When we examine the ingredients that make up the individual pastes, we notice that many of them were considered to possess aromatic qualitities.[138] A correspondence between early Buddhist and āyurvedic medicine is therefore implied.

The name of the affliction given in the Pāli texts is significant. The term *chavi* is an old Sanskrit (Vedic) term for "skin," which was replaced in classical Sanskrit by *tvac*. Pāli *dosa* and Sanskrit *doṣa* have the principal meaning "corruption," referring in the technical sense to a corruption of the primary elements (*dhātu*)—that is, "'peccant' humors." The compound *chavidosa*, therefore, is synonymous with *tvagodṣa*, an expression that included *kuṣṭha* used predominantly by Suśruta.[139] The causes for *tvagdoṣa*, as outlined by Suśruta, have nothing to do with the humors.[140] They included the imporper intake of food, suppression of natural urges, improper use of medicated oil, and bad actions. The compound *tvagdoṣa* (= *chavidosa*), therefore, appears to be an older generic expression for a corruption of the skin or skin disease, which included *kuṣṭha*.[141] Therapies for treating skin disease in both traditions derived from a common source of medical lore and later were elaborated by the medical intellectuals.

Body Filled (with the "Peccant" Humors)

A certain monk suffered from having his body filled (*abhisannakāya*), according to Buddhaghosa, with the "peccant" humors (*dosas*). The treatment for this affliction involved the drinking of purgative (*virecanaṃ pātum*) and the taking of (1) clarified sour rice gruel (*acchakañjiya*); (2) natural (i.e., unprepared) soup (*akaṭayūsa*), which Buddhaghosa describes as a nonoily drink cooked with green gram (*mugga*, Skt. *mudga*);[142] (3) both unnatural (i.e., prepared) and natural (i.e., unprepared) (soup) (*kaṭākaṭa*), which, according to Buddhaghosa, is only slightly oily; and (4) meat broth (*paṭicchādaniya*), which, Buddhaghosa says, has the flavor of flesh.[143]

For the first time, the notion of "'peccant' humor" (*doṣa*, Pāli *dosa*) is implied in Buddhist monastic medicine, hinting at a more specialized meaning corresponding perhaps to that of *doṣa* in the medical treatises.

Caraka uses a phrase similar to that found in the Pāli to speak generally of humoral disease: *doṣair abhikhinnaś ca yo naraḥ* (and a man who is afflicted with the "peccant" humors).[144] A variant reading for this may help to elucidate its more exact meaning: *doṣair abhiṣyannaś ca yo naraḥ* (and a man who is oozing [or moistened] with the "peccant" humors).

Suśruta begins his chapter on "the treatment of supervenient diseases cured by emetics (*vamana*) and purgatives (*virecana*)" with the following axiom: "The diminished humors [*doṣa*s] are to be strengthened, the provoked are to be pacified, and the increased are to be diminished; they are to be maintained in equilibrium." He goes on to state that "emetics and purgatives are used principally to remove the humors" and outlines the method by which this is accomplished: a few days prior to receiving an emetic or a purgative, the patient is oiled and sweated. Just before a purgative is administered, the patient should eat light food and drink hot water and sour fruits, and on the next day he is given the purgative to drink. After a proper purgation, the patient may drink a light and lukewarm thin gruel (*peyā*).[145] This, in broad outline, is the course of treatment given to the sick monk in the Buddhist canonical account.

Examination of the chapters on emetics and purgatives in the early medical treatises indicates that both emetics and purgatives are mentioned together and that the drugs and procedures for one often are, with minor modifications, used for the other.[146] The initial techniques of oiling and sudation and the giving of light food are the same for emetics and purgatives. Suśruta states that green gram (*mudga*) soup (*yūṣa*), saturated with the essence of purgative drugs and combined with clarified butter and rock salt, is a good purgative. Soups with other leguminous grains may be used in the same way; and emetic medicines may be given by means of soup.[147] A medium of soup, therefore, was commonly used to administer both purgatives and emetics. After an emetic, the patient should consume a soup (*yūṣa*) made, among other things, from green gram (*mudga*) or the meat of jungle animals.[148] This differs slightly from the lukewarm gruel (*peyā*) given after purgation but seems to be closer to the Buddhist's meat both. The variation in the postevacuative diet is only minor and may well reflect the interchangeability between emetic and purgative procedures.

Based on the medical texts, we may tentatively understand the use of the four liquids in the Buddhist canonical account as follows:

1. The sour gruel was prepurgative food.
2. The natural soup, which the commentator says is a nonoily drink cooked with green gram (*mudga*), and the natural and unnatural

soup, which the commentator explains as slightly oily, are mediums for the purgative and means for oiling the body that is part of the purgative treatment.

3. The meat broth is the postpurgative diet.

The use of oil would have been appropriate for an ascetic with a mild constitution.

From the preceding analysis, it appears that some notion of the "peccant" humors (*doṣa*s) and their effects on the health of the body was current at the time the Vinaya rules were compiled. There is also a definite continuity between early Buddhist and early āyurvedic medicine with respect to the treatment of a condition of excessive bodily humors by means of evacuation. The methods of administering the emetics and purgatives are remarkably similar in the texts of the early medical authors, reflecting a common source of which the Buddhists also had knowledge.

Wind in the Abdomen

The affliction of wind in the abdomen (*udaravāta*) is discussed two times in the Vinaya. In the first case, a monk was cured by being given a mixture of salt and a type of astringent barley wine (*loṇasovīraka*) to drink. This treatment was allowed to the sick (*gilāna*), but for those not sick (*agilāna*), it was permitted as a drink (*pāna*) when mixed with water (*udakasambhinna*).[149]

The same treatment is mentioned elsewhere is the Vinaya, where the Pāli is *loṇasuvīraka*.[150] Here Buddhaghosa defines it as "a medicine made from all the tastes" and details its preparation: it is composed of the three myrobalans (*triphalā*), all the grains and cooked cereals, gruel, all fruits beinning with the fruit of the plantain, all top sprouts begining with the sprouts of the fragrant screwpine (*ketaka*) and the wild date (?) (*khajjūtī̌*), numerous bits of fish meat, honey, molasses, rock salt, and medicines beginning with the bitters. It should be stored in a pot, the rim of which has been greased, and set aside for one, two, or three years. When it is ripe, it is the color of blackberry (*jambura*) juice. Among other things, it is a good medicine against wind (*vāta*), cough (*kāsa*), skin disease (*kuṭṭha*), morbid pallor or jaundice (*paṇḍu*), and fistulas (*bhagandala*). It should be taken in its natural state by the sick (*gilāna*) and mixed with water by those not sick (*agilāna*).[151]

In the second case, the Buddha suffered from wind in the abdomen (*udaravāta*). His treatment, however, was different. He was given to drink a rice gruel (*yāgu*) containing three pungent substances (*tekaṭula*)—sesame (*tila*), rice grain (*taṇḍula*), and green gram (*mugga*)—which were previously prepared, purified (or stored: *vāsetvā*),[152] and cooked. The text states that this treatment had been given to him before for the same ailment.[153] This medicine is also mentioned elsewhere in the Vinaya and is explained there by Buddhaghosa, who states that it might be composed of the three ingredients: sesame (*tila*), rice grain (*taṇḍula*), and green gram (*mugga*). However, black gram (*māsa*, Skt. *māṣa*), horse gram (*kulattha*), or any single cooked cereal together with sesame (*tila*) and rice grain (*taṇḍula*) could be used to make the three. It is then prepared by mixing the three in milk and four portions of water, together with ingredients beginning with clarified butter, honey, and (granulated) sugar.[154]

Each of the early medical treatises devotes a chapter or two to the causes and treatments of diseases of the abdomen (*udara*).[155] Of the eight types of abdominal disorders, there are those caused by the humors individually and collectively. The one caused by the humor wind (*vāta*) is called *vātodara* (wind in the abdomen), which could correspond to the Pāli *udaravāta*. The treatment for this disorder, however, differs from that outlined in the Pāli sources. If the patient is strong, he should be given an oil treatment, sweated, and purged. The abdomen should be wrapped to prevent its increase by wind. He should be purged daily and given milk to drink, and if the wind begins to move upward, the digestive fire should be activated by the intake of slightly soured and salted vegetable or meat soup (*yūṣai rasair vā mandāmlalavaṇair*); the patient should again be oiled and sweated, and a nonoily enema given to him. If the patient is weak, old, or of a delicate constitution with little accumulation of wind, the physician should treat him with soothing measures (*śamana*), which consist of clarified butter, vegetable or meat soup, rice, enemas (*vasti*), massage (*abhyaṅga*) and oily enemas (*anuvāsa*), and the intake of milk.[156]

Neither the two treatments mentioned in the Buddhist canonical texts nor the further details about them elaborated by Buddhaghosa correspond to the measures prescribed by the medical authors in the case of wind in the abdomen.

Caraka provides one treatment that could be used for the elimination of the humors in all types of abdominal disorders. It is a type of rice gruel (*yavāgū*, Pāli *yāgu*), which bears a slight resemblance to that given to the Buddha: the rice gruel should be consumed with vegetable or meat soup, containing slightly sour, fatty, and pungent (*kaṭu*) ingredients and cooked with the five medicinal root plants (*pañcamūla*).[157]

Sanskrit *sauvīraka* (Pāli *sovīraka*, *suvīraka*) is a type of barely wine, combined with clarified butter and other grains and plants. Caraka mentions it as a purgative and gives the following two recipes for it: "*Kulmāṣa* [Cakrapāṇdatta 'boiled barely cake'], with barely boiled with a decoction of black turpeth tree [*śyāmā*] and [red-rooted] turpeth tree [*trivṛt*], is fermented with water for six days in a pile of grains";[158] and "One should ferment fried barely grains [*maruja*] in water boiled with 'ram's horn' [*meṣaśṛṅgī*], yellow myrobalan [*abhayā*], long pepper [*kṛṣṇā*], and leadwort [*citraka*]. And when that *sauvīraka* is produced,..."[159] Suśruta gives a more detailed recipe for its preparation:

> Roots beginning with [red-rooted] turpeth [*trivṛt*], of the first group,[160] of the great group of five roots, [161] and of both bowstring hemp [*mūrvā*] and jequirity [?] [*śārṅgaṣṭā*];[162] [along with] common milkhedge [*sughā*], sweet flag [?] [*haimavatī*], three myrobalans [*triphalā*], Indian atees [*ativiṣā*], and sweet flag [*vacā*]. Having combined them, one should divide them into two [equal] portions. One of them should be made into a decoction [*niḥkvātha*]; the other one, into a powder [*cūrṇa*]. Pulverized barely should often be [placed] in this decoction [Ḍalhaṇa: "It should be in it for a period of seven days"]. It is considered proper to have these portions of [the barely] dried and lightly baked. Having [then] taken the fourth part of powder, mentioned here, the entire [mixture of powder and barely] should thereupon be tossed into a pot [*kalaśa*]. Then, well supplied with the cooled decoction of those [drugs beginning with (red-rooted) turpeth (*trivṛt*) (Ḍalhaṇa)], [the solution] should be deposited [in a pot for fermentation] as before. For this is known as *sauvīraka*.[163]

A considerable difference exists between the recipe given by Buddhaghosa and those found in the medical texts; however, correspondence occurs in the ingredient of three myobalans (*triphalā*) and with the fermentation in a pot found in Buddhaghosa and Suśruta. Clearly, different methods for preparing the barely wine existed in ancient India.

Elsewhere in medical treatises, *sauvīraka* is described as a sour appetizer and is included with other drugs as a general treatment for various ailments and as a purgative.[164] It was a very common form of spirituous drug.

The use of salt (Skt. *lavaṇa*, Pāli *loṇa*) figures in several of the treatments for abdominal disorders outlined in the works of Caraka and Suśruta. In Suśruta, salt and *sauvīraka* are used with other drugs to soothe a rectal fistula (*bhagandara*), to cure internal (abdominal) tumor (*gulma*) caused by wind, and to create an appetite in the patient.[165] In Caraka, inhabitants of the *sauvīra* region are said to be excessive users of salt (*lavaṇa*). As a result, they suffer physical defects. Hence, too much salt should be avoided.[166] The *sauvīra* region was probably located in northern India, and *sauvīra* referred to the people of that area. It is possible that they developed the barely wine that bears their name and employed it with salt in the treatment of abdominal disorders. Wandering ascetics could have acquired the technique from these people and, by their transmission, contributed to its incorporation in Buddhist monastic medicine and in the teachings of the early medical tradition.

The morbid condition of wind in the abdomen is expressed similarly in early Buddhist and early āyurvedic medicine. The treatments for this disease, however, do not correspond. Parts of the early Buddhist treatments and the later elaboration of them by Buddhaghosa are found in the appropriate chapters of the medical treatises, but exact parallels are wanting. The existence of such significant differences suggests that a variety of treatments and recipes existed under the same name. The use of *sauvīraka* with salt for abdominal disorders caused by wind is found in the medical tradition of Suśruta. This points to a correspondence between early Buddhist and āyurvedic medicine and reflects a common source. The similarities that do occur, however, are found in general treatments of abdominal disorders rather than in specific cures of wind in the abdomen. This points to a common storehouse of medical lore concerning abdominal diseases from which the early Buddhist monastic and early āyurvedic medicinal traditions derived their information. Variation in treatments of specific disorders probably resulted from local practices that necessitated substitutions and modifications based on the availability of drugs. Moreover, the general use of *sauvīraka* barley wine as a medicine may well have derived from the people of the region that bears its name (i.e., *sauvīra*). Its use with salt was perhaps a specific treatment developed by these people, picked up and assimilated by wandering ascetics, including Buddhists.

Burning in the Body

The monk Sāriputta suffered from the affliction of burning in the body (*kāyaḍāhābādha*). He was treated with lotus sprouts (*bhisa*) and stalks (*muḷālikā*), bringing the burning in his body to an end.[167] The brevity of this case inhibits a precise determination of the disease from which the monk suffered. Certain similarities among the medical texts permit at least a plausible connection.

In the early medical treatises, a burning sensation of the body (*dāha*) is generally a symptom of fever (*jvara*),[168] and various types and parts of lotuses are often employed in treatment to remove the bodily heat.[169] One therapy required the patient to lie down on a bed covered with the cooling shoots (*dala*) of species of the lotus plant (*puṣkara*, *padma*, *utpala*), leaves of the plantain, and other cooling elements. He should be cooled by fanning with species of the lotus and sprinkled with water steeped with sandalwood, and should be bathed in rivers, pools, and lakes with lotuses and clean water.[170]

Suśruta does not mention such a treatment in his chapter on fevers (*jvara*) but prescribes that various species of lotus be used in a bath for the removal of burning sensations in the body.[171] Elsewhere in the *Suśruta Saṃhitā*, excessive bodily heat (*dāha*) appears as a separate disease unconnected with fever (*jvara*). A therapy for this morbid condition resembles that outlined in the *Caraka Saṃhitā* for a burning body caused by fever: the patient should lie down on a bed of moistened lotus flowers and leaves and become cooled by being surrounded and imbued with the sight, touch, and scent of lotuses. This remedy is prescribed specifically for a burning body resulting from vitiated bile caused by intoxication (*pāna*) and generally in cases of hot bodily sensations caused by hemorrhagic disorders (*raktapitta*) and thirst (*tṛṣ*).[172]

Use of different parts and types of the lotus for the purpose of cooling a patient whose body is racked by extreme heat forms part of the early medical traditions of both Caraka and Suśruta.[173] Suśruta's inclusion of therapeutic employment of lotuses against a specific condition of bodily sensations of heat, unrelated to fever, points to a derivative use of cooling lotus remedies based on a subsequent classification of fever symptoms into separate diseases. Their primary use was as a treatment for fever, which is supported by their application in early Buddhist medicine for a febrile condition exemplified by burning over the entire body. Although not stated, Sāriputta's remedy, like that in the medical treatises, probably involved his lying on a bed of lotus sprouts and stalks in order to cool his body. The use of lotuses in the treatment of fevers, therefore, likely derived from a source common both to early Buddhist medicine and to the tradition of Caraka, whose description of their use suggests how they might have been employed by the Buddhist monks.

Rectal Fistula

This is a case of a proscribed rather than prescribed therapy. A monk suffered from a rectal fistula (*bhagandala*), and the physician (*vejja*) Ākāsagotta of Rājagaha lanced it with a knife (*satthakamma*). The opened fistula, having the appearance of a varan's or monitor lizard's mouth (*godhāmukha*), was then shown to the Buddha, who objected to such a method of treatment, saying that the skin is too tender at the private parts, the wound hard to heal, and the knife difficult to guide. As an alternative to the use of the knife, a group of six monks suggested that an enema treatment (*vatthikamma*) be administered. The Buddha again rejected the therapy and proclaimed the rule that neither a treatment with a knife within two finger breadths of the private parts nor an enema treatment is to be carried out. Buddhaghosa expands and elaborates on this as follows: every treatment with the knife involving cutting off, splitting, piercing, or scratching with a knife, thread, thorn dagger (?) (*santikā*, var. *sattikā*), splinter of stone, or thorn is surely not to be done. Any enema treatment utilizing animal skin or cloth and involving potential injury to the bladder is surely not to be carried out. The treatment with the knife is not to be undertaken within two finger breadths of the private parts, and the enema therapy is forbidden in the pudenda. A type of suppository smeared with medicine or a stalk of bamboo by which an alkaline treatment is administered or oil introduced into the rectum is the proper mode of treatment.[174]

The early medical treatises mention the disease *bhagandara* (Pāli *bhagandala*) (rectal fistula). Suśruta devotes two chapters to its causes and cures.[175] He defines the *bhagandara* as eruptions (*piḍakā*) that have suppurated.[176] Caraka provides a similar definition and recommends the following course of treatment: purgation, probing (*eṣaṇa*), incision (*pāṭana*), cauterization with (hot ?) oil (*tailadāha*) after cleaning the path, and treatment with a well-prepared (i.e., cooked) caustic thread (*kṣārasūtra*). Suśruta's course of treatment follows that of Caraka but stresses the more surgical procedures: the probe (*eṣaṇī*) is applied to expose the mouth of the fistula, which is opened with a knife (*śastra*). He also prescribes cauterization with fire (*agni*) or with caustics (*kṣāra*).[177] Different types of medicines combined with sesame oil are applied to the fistula to purify and close the opened wound.[178] Suśruta states that the remedial procedures given for the two types of wounds (*dvivraṇa*; i.e., those caused by internal means and those brought about by external causes) should also be followed in the treatment of fistulas.[179] Purgation, probing, incisions, use of caustics, bloodletting, and enema treatment (*vastikarman*) are among the treatments prescribed for the two types of wound.[180]

The case mentioned in the Vinaya, requiring the use of a knife or, alternatively, enema treatment, follows closely that prescribed in the early medical treatises and is favored by Suśruta, suggesting that the physician Ākāsagotta may well have been a follower of the tradition of Suśruta or Dhanvantari. A prohibition against such treatments is wanting in the medical books. Buddhaghosa's comments indicate that by the fifth century C.E. an effective substitute therapy, finding a parallel in the medical treatises, had become part of Buddhist monastic medicine.

Evidence points to lancing or enemas as the predominant means for treating suppurating rectal fistulas in ancient India. The Buddhist prohibition against cutting near the private parts and the application of enemas may well have given rise to alternative modes of treatment involving the application of (caustic) medicines with bamboo splinters, which were eventually incorporated in the early medical textbooks. A distinction between traditional schools of surgery and of internal medicine is noticed in the Pāli sources. Each existed alongside the other and contributed to the storehouse of India's ancient medical lore. The occurrence of medical practitioners in the *saṅgha* establishes the close association between the Buddhist monastery and medicine, and the Buddhist proscription of existing therapies for fistulas indicates the important role played by Buddhism in bringing about new medical treatments that became part of the medical tradition.

Conclusion

The purpose of our study was the investigation of Indian medicine in the crucial but neglected period from about 800 to 100 B.C.E. in order to obtain a more comprehensive and more plausible picture of ancient Indian medical history than had previously been presented. Emphasis was placed on the ascertainment of the socioreligious dynamics that brought about the change in the conceptual basis of ancient Indian medicine during this period, which saw the transition from Vedic medicine, anchored in a magico-religious ideology, to *āyurveda*, dominated by an empirico-rational epistemology. The transition was conceived in terms of a paradigm shift caused not by an orthodox brāhmaṇic but by a heterodox renunciant intellectual tradition unencumbered by the socioreligious ideology of brāhmaṇic orthodoxy. It was acknowledged that the Indian example does not fit precisely Thomas Kuhn's theory of paradigm shifts because of the incorporation of previous magico-religious healing techniques, which performed a legitimizing function by providing continuity with the past, into an entirely new empirico-rational approach to disease and its cure. Nevertheless, in all other respects the radical change in the conceptual framework of medicine closely approximates Kuhn's model, which is useful in emphasizing the importance of this revolutionary change. The methodology required a historical and philological investigation of a variety of Indian and non-Indian sources, with special emphasis on Buddhist materials, for they were a virtually untapped mine of valuable information pertinent to the topic.

The results of our study present rather unconventional findings, which the traditionalist may find offensive. The picture of the evolution of India's antique medical heritage is very different from that portrayed by the teachers and authors of *āyurveda*, who represent traditional Indian medicine as a brāhmaṇic science from its inception. A critical analysis of the sources demonstrates that this viewpoint results from Hindu intellectual endeavors to render a fundamentally heterodox science orthodox. From the early Vedic period, medicine and healers were excluded from the core of the orthodox brāhmaṇic social and religious hierarchy and found acceptance among the heterodox traditions of mendicant ascetics, or *śramaṇa*s, who became the repository of a vast storehouse of medical knowledge. Unaffected by brāhmaṇic strictures and taboos, these śramaṇic physicians developed an empirically based medical epistemology and accumulated medical lore from different healing traditions in ancient India. Ideally suited to the Buddha's key teaching of the Middle Way, this medical information was codified in the early Buddhist monastic rules, which stressed the practical rather than the theoretical virtues of healing and gave rise to a tradition of Buddhist monastic medicine during the centuries following the founder's death.

Buddhism played a key role in the advancement of Indian medicine through its institutionalization of medicine in the Buddhist monastery. The medical doctrines codified in the monastic rules probably provided the literary model for the subsequent enchiridions of medical practice, gave rise to monk-healers and to the establishment of monastic hospices and infirmaries, and proved to be beneficial assets in the diffusion of Buddhism throughout the subcontinent during and after the time of Aśoka. The close connection between healing and Buddhist monasticism eventually led to the teaching of medicine as one of the five sciences (*vidyā*s) in the large conglomerate monasteries of the Gupta period, and the healing arts, adapted to fit appropriate social and cultural settings, went along with Buddhism as it spread to other parts of Asia. Hinduism assimilated the ascetic medical repository into its socioreligious and intellectual tradition beginning probably during the Gupta period and by the application of a brāhmaṇic veneer made it an orthodox Hindu science. The earliest extant medical treatises, the *Caraka* and *Suśruta Saṃhitā*s, bear distinctive indications of this Hinduization process. Hindu monastic institutions also followed the Buddhist model and established infirmaries, hospices, and eventually hospitals in their monasteries.

Knowledge of the process by which Buddhism integrated medicine into its religious doctrines and practices not only helps to deepen the understanding of Indian medical history, but also sheds abundant light on the social history of Buddhism through the centuries. The ideal blending of medicine and religion in Buddhism remains an essential aspect of its religious tradition in the modern era, and consideration of the role of healing is an essential but frequently neglected component of any complete investigation or description of current Asian Buddhism.

Although a clear picture of the evolution of Indian medicine emerges from a critical examination of the relevant sources, an important piece of the puzzle remains missing and constitutes a principal topic for further investigation. Central to *āyurveda* is an etiology based on three "peccant" humors, or *doṣa*s: wind, bile, and phlegm. Evidence of a complete formulation of this theory can be traced in Buddhist Pāli literature, where the three humors and their combination are mentioned as causes of disease. Unconvincing attempts to find origins of the theory in the Vedic *Saṃhitā*s and *Brāhmaṇa*s suggest that the Buddhist references are perhaps the earliest attested citations of the humoral etiology outside the classical treatises of Caraka and Suśruta. Indian medical epistemology based on empiricism is traceable to the śramaṇic traditions, but evidence of an evolving theory of disease causation derived from empirical data is wanting. A crucial link between empiricism and theory is missing. Further research is required to determine whence the theory derived and how and when it became part of the medical tradition. The tendency of many scholars to view it as somehow connected to the four-humor theory of Hellenistic medicine is untenable because the humors of Greco-Roman medicine included, in addition to wind, two forms of bile (black and yellow) rather than the single bile of Indian medicine, and the fourth, blood, is not listed as an Indian humor. Although the numerology of the humors may not correspond, the fundamental idea of attributing disease to a corruption of basic elements in the body remains common to Hellenistic, Indian, and Chinese medical theories. Given the peripatetic life-style and empirical orientation of the śramaṇic physicians, contact with travelers bearing medical knowledge from distant lands was a distinct possibility. The medical information exchanged and subsequently adapted by the Indian physicians could have resulted in the threefold humoral etiology characteristic of *āyurveda.* Against the background of India's śramaṇic medical heritage provided in this study, critical examination of Hellenistic, Chinese, and Indian sources, including the literature of the śramaṇic Jainas, should deepen our understanding of ancient Indian medical history by supplying not only its missing theoretical link but a wealth of new information pertinent to the repository of medical knowledge from which both āyurvedic and Buddhist monastic medicine derived their traditions. Similarly, an investigation of the connection between the Yogic and the medical traditions would help elucidate further aspects of asceticism and healing in ancient India.

APPENDIX I

Jīvaka's Cures

Although not part of Buddhist monastic medicine, the cures performed by the physician Jīvaka Komārabhacca elucidate the medical doctrines and practices present during the centuries before the common era and further demonstrate the close association between the early Buddhist monastery and the healing arts. As noted in Chapter 4, the episodes became a popular Buddhist folk legend transmitted with appropriate modifications to Asian cultures outside India with the diffusion of Buddhism and thus helped to establish medicine as an integral part of the religious tradition.

In an article published in 1982, I analyzed Jīvaka's cures in relationship to corresponding therapeutics explained in the early medical treatises of Caraka and Suśruta.[1] Subsequent research uncovered additional information revealing trends in ancient Indian medical lore and the connection between Buddhism and the medical tradition. The appendix therefore contains a brief survey of Jīvaka's cures incorporating recent data.

Disease of the Head

Two cases of head disease are mentioned in the monastic code and involve the physician Jīvaka Komārabhacca. The first was a persistent head affliction (*sīsābādha*) lasting seven years, suffered by a Sāketin merchant's wife. The treatment, administered by Jīvaka, was nasal therapy (*natthukamma*). A handful of clarified butter (*sappi*), decocted with various medicines (*nānābhesajja*), was given through the nose of the woman, who was supine on a bed. The clarified butter then issued from the mouth and was spat out into a receptacle.[2]

The second case was a head affliction (*sīsābādha*), also lasting seven years, suffered by a merchant from Rājagaha. The cause of the affliction is stated as being two living creatures (*dve pāṇake*) residing in the skull. Jīvaka treated this patient surgically by cutting away the skin of the head of the merchant, who, supine on a bed, was bound to it. Jīvaka proceeded to twist open a suture and remove a large and a small creature. The suture was then closed, the skin sewn back, and a medicated paste (*ālepa*) applied.[3]

These two cases offer important information about Buddhist medical knowledge and its sources in relationship to the early āyurvedic medical tradition. The treatment of the merchant's wife by means of nasal therapy corresponds closely to the prescribed course of action against head disease (*śiroroga*) by means of nasal therapy (*nastakarman*) outlined in the medical treatises of Caraka and Suśruta.[4] The treatment of the merchant, involving a type of trepanation, however, finds no parallel in the early medical literature. Archaeological evidence demonstrates that trepanation was practiced in parts of ancient northwestern India, indicating a medical tradition whose use of trepanation was not incorporated in the early āyurvedic medical compilations.[5]

Although not contained in the *Bhesajjakkhandhaka*, characterizing the Buddhist monastic medical tradition, these two cases of head affliction preserved in the Vinaya provide clear evidence of different healing traditions in ancient India. Viewed against the background of Buddhist medicine previously examined, they offer further support to the claim that the storehouse of medical knowledge that gave rise to Buddhist monastic medicine both contributed to the medical doctrine contained in the treatises of Caraka (Bhela) and Suśruta and was not codified in the extant medical sources. These stories therefore provide compelling proof that a plurality of medical traditions constituted the common storehouse of ancient Indian medical lore.

Rectal Fistula

Jīvaka cured King Seniya Bimbisāra of Magadha of a rectal fistula (*bhagandala*) by means of the application of a medicated ointment or salve, administered with his fingernail.[6] The medical treatises present various causes and remedies of rectal fistula (*bhagandara*). The therapeutics include purgation, probing, incision, cauterization with fire or caustics, lancing, bloodletting, and enema therapy.[7] The specific remedy employed by Jīvaka does not occur in the medical texts. The therapeutic procedure utilizing caustic medicines, however, might bear a similarity to this course of action. In the āyurvedic treatises, probes (*eṣaṇī*), sometimes called *śalākā* in the later texts, were often used to apply caustics to boils and wounds. The *śalākā*s were of three types, resembling, respectively, the nails of the small, ring, and middle fingers. The *śalākā* (Pāli *salākā*) in Buddhist medicine is defined as a bamboo splinter with caustic medicines used in the treatment of boils and wounds. By analogy, then, the fingernail could have served the same function as the *śalākā* in the treatment of a rectal fistula.[8] Use of a bamboo splinter to apply caustic remedies in the treatment of a rectal fistula was unknown to Buddhist monastic medicine until the fifth century C.E., when Buddhaghosa introduced the procedure as an effective substitute for lancing and enema therapy.[9] The technique of administering medicines with a bamboo splinter or a splinterlike fingernail occurs both in the account of Jīvaka's cure and in early āyurvedic therapeutics, suggesting a common source for the remedy, which might have originated in Buddhist circles.

Knot in the Bowels

A certain merchant's son from Vārāṇasī suffered from the affliction of a knot in the bowels (*antagaṇṭhābādha*, var. *antagaṇḍābādha* [affliction of swelling in the bowels]), resulting from turning somersaults with a stick. It prevented him from digesting food and evacuating feces and urine. The treatment involved a type of laparotomy performed by the physician Jīvaka. The patient was first bound to a post; then Jīvaka cut into the abdomen, extracted the knotted bowels, disentangled them, replaced the corrected bowels, sewed up the incision, and applied a medicated salve (*ālepa*), resulting in the merchant's son's full recovery.[10]

An exact equivalent to the Pāli *antagaṇṭha* (var. *antagaṇḍa*) is wanting in the early medical literature. However, two afflictions involving injury to the bowels (*antra*, Pāli *anta*) are found in the treatise of Suśruta. An enlargement of the bowels (*antravṛddhi*) is described by Suśruta as follows:

> The wind, increased and enraged by the types of exertion beginning with the carrying of a heavy load, the engaging in conflict with some powerful person, or the falling out of a tree, separates a part of the bowels[11] into two parts [Gayadāsa]; [and], having gone down and reached the conjunction of the groin [i.e., scrotum], [the wind] becomes situated [there] in the form of a knot [*granthi*]. In the incurable type, [the wind], having entered the scrotum, causes pain after some time. When [the area with the wind] becomes very extended [and] swollen like a bladder, pain occurs; when pressed downward and upward, it makes the same sound;[12] when it is let loose, it inflates again.[13]

Suśruta's description seems to be of a type of hernia, brought on by physical exertion, similar to the activity engaged in by the youth in the Buddhist account. Suśruta's understanding that the enlarged bowel was caused by wind (*vāyu*), which separates the bowels (*antra*) and situates in the scrotum like a knot, may well be a more detailed and technical explanation of the affliction suffered by the merchant's son. Any hint of these details, however, is wanting in the Buddhist account.

Suśruta also details the treatment of *antravṛddhi*, explaining that when the knot reaches the scrotum it is incurable and treatment should be abandoned. When it has not yet reached the scrotum, however, it should be treated by the method of *vātavṛddhi*[14] and cauterized. Bloodletting is also recommended as part of the therapy.[15] This course of action does not find a parallel in the Buddhist story.

A second affliction involving the bowels is again found in Suśruta, with a variant in Caraka. It is a type of wound from which the unbroken part of the bowel (*antra*) has protruded. The treatment for this condition required that the part of the bowel be washed with milk, lubricated with clarified butter, and gently replaced into its original position. Surgery was used when the reintroduction was made difficult by the size of the outer wound. In all cases, the wound or incision was to be sewn and a medicated oil, prepared with different barks and roots, was to be placed on the wound to promote its healing. This condition outlined by Suśruta has no technical name. Its description is that of a wound to the abdomen, from which a part of the bowel protrudes. No causes are offered. The treatment required the replacing of the exposed bowel, using surgery if necessary, and the application of medicinal drugs on the open wound.[16] The entire procedure bears only a slight similarity to that treatment performed by Jīvaka.

Neither of these two afflictions given in the treatise of Suśruta is exactly parallel to the Buddhist account of the son's twisted bowels. The Sanskrit *antravṛddhi* (enlargement of the bowels) is close to the Pāli *antagaṇṭha*. Suśruta's course of treatment for this condition, however, bears no similarity to that executed by Jīvaka. Jyotir Mitra suggests that the affliction might be compared with a symptom of an abdominal obstruction (*baddhagudodara*) known as *antrasammūrcchana* (swelling of the bowels) detailed by Caraka. The commentator Cakrapāṇidatta glosses this compound as "turning round of the bowels" (*antraparivartana*), which Mitra translates as "entanglement of intestines."[17] The symptoms of this abdominal obstruction include those suffered by the merchant's son. The treatment of the condition involved the opening of the abdomen, examination of the bowels for obstructions or perforations, removal of any obstruction or repair of any perforation by the use of ants, replacing of the bowels, and suturing of the wound to the abdominal wall. Because of these coincidences between the Vinaya passage and the statements in the medical treatise of Caraka, Mitra concludes that the boy suffered from an abdominal obstruction.[18]

Like Suśruta's two conditions and treatments outlined above, Caraka's reference offered by Mitra does not linguistically parallel the Pāli *antagaṇṭha*. Cakrapāṇidatta's gloss brings it closer, but a precise equation is wanting. Of the three possible explanations presented above, that found in Caraka appears to be the nearest to the Pāli. The Pāli variant, *antagaṇḍa* (swelling in the bowels), may therefore be justified if the condition was a form of abdominal obstruction, causing the bowels to protrude.

The Buddhist author was describing an injury to the bowels caused by acrobatics and its cure by means of surgery. An exact corresponding form of the affliction is wanting in the earliest extant medical literature; but similar ailments are covered in detail, and the treatment for them bears resemblances to that performed by the physician Jīvaka. The similarities therefore point to a common source of medical knowledge and a continuity of medical doctrine.

Morbid Pallor or Jaundice

King Pajjota of Ujjenī suffered from the affliction of morbid pallor (*paṇḍuroga*). Jīvaka cured him with a surreptitious application of a decoction (*kasāva*) of clarified butter (*sappi*) cooked with various medicines (*nānābhesajja*) in order to make the king vomit.[19]

Both the *Caraka* and *Suśruta Saṃhitā*s prescribe the administration of an evacuative using clarified butter decocted with other medicines as a principal remedy for a patient suffering from morbid pallor (*pāṇḍuroga*).[20] Caraka specifically mentions the uses of the decoction *pathyāghṛta* (yellow myrobalan [*harītakī*] cooked with clarified butter),[21] and Suśruta explains the typical administration of clarified butter along with drugs such as yellow myrobalan (*harītakī*), the three myrobalans (*triphalā*), and the lodh tree (*tilvaka*):

> Having carefully examined the one afflicted with morbid pallor [and having known him to be] curable, the [physician] should oil [the patient] with clarified butter [*ghṛta*] and should cleanse [i.e., evacuate] [him] from top to bottom by the use of powdered *harītakī* mixed with honey and clarified butter; either [the patient] should drink the clarified butter cooked with turmeric [*rajanī*] or with *triphalā* or *tilvaka*,[22] or he should drink [a substance] made from purgative drugs and purgative drugs mixed with clarified butter.[23]

Jīvaka's remedy and a principal treatment for morbid pallor prescribed in the early medical treatises clearly derived from the same source.

Body Filled with the "Peccant" Humors

The Buddha suffered from a condition in which his body was filled with the "peccant" humors (*kāya dosābhisanna*). Jīvaka attended to him and administered a mild purgative. The treatment involved first oiling the body for a few days (*kāyaṃ katipāhaṃ sinehetham*), followed by the inhalation (*upasiṅghatu*) of the fragrance of three individual handfuls of lotuses mixed with various medicines, and finally a bath (*nahāta*), resulting in thirty purges. The Buddha was then instructed to eat only soup (*yūsa*) until his body returned to a normal condition.[24]

In another case, monks became very ill (*bahvābādha*), with their bodies filled with the "peccant" humors (*abhisannakāya*), after having eaten sumptuous foods (*paṇitāni bhojanāni*). Jīvaka, in this instance, recommended that the monks walk off the ill effects of overindulgence.[25]

In both cases, the particular malady from which the patients suffered was a body full of humors, which, in at least one instance, was known to be caused by overeating. The common treatment required purgation, followed by light food. Exercise was also an alternative remedy, especially in the case of the consumption of unsuitable food.

Caraka defines a patient suffering from a body filled with the "peccant" humors in terms similar to the Pāli account (i.e., *doṣair abhikhinnaś ca yo naraḥ* [and a man who is afflicted with the "peccant" humors]),[26] and Suśruta outlines the basic course of treatment: Prior to receiving an evacuative, the patient is oiled and sweated and given light food, hot water, and sour fruits to eat on the day before the treatment. After purgation, the patient should take a light and lukewarm, thin gruel, allowing the bodily constitution to normalize itself.[27] This corresponds broadly to Jīvaka's treatment of the Buddha.

In the case of a patient who is very delicate (i.e., who has a delicate constitution) or who is averse to drugs, the medical texts prescribe a special type of emetic. The powder of the seeds of the emetic nut (*madana*) or of the ribbed luffa (*kṛtavedhana*), repeatedly soaked in a decoction of emetic drugs such as the emetic nut (*madana*) should be sprinkled over large lotus flowers (Ca: *bṛhatsaroruha*) or other lotus flowers (Su: *utpalādi*). The patient should then smell the flowers and, by inhaling the powder, throw up. Suśruta adds: "Likewise, [the physician] should administer this [treatment] in the case of those whose humors are deranged (*anavabaddhadoṣa*) and in the case of those who have drunk barley gruel up to [their] throats [i.e., up to their capacity]."[28] This is remarkably similar to the evacuative that Jīvaka gave to the Buddha.

A variant of the special treatment is found as a *virecana* (purgative) in Suśruta:

> Having smelled the flowered garlands scattered with the powder of golden cleome [*saptalā*], canscora [*śaṅkhinī*], wild croton [*dantī*], (red-rooted) turpeth [*trivṛt*], and purging cassia [*āragvadha*], after having been soaked in cow's urine and in the milky sap of the milk hedge tree [*snuhī*] for a week, or having put on clothes scattered with this powder, a person with a delicate constitution [*mṛdukoṣṭha*] is duly purged.[29]

Caraka prescribes that equal quantities of these drugs should be soaked in cow's urine overnight and dried in the sun the next day. After this is done for a period of seven days, the drugs should be soaked in the milky sap of the milk hedge tree for a week. A powder made from the substance should be put on a garland or on cloths. The inhalation of the powder produces an easy purgation for kings with delicate constitutions.[30]

Although the use of a lotus flower is not mentioned in this form of purgation, the method by which the medicine is administered to the patient with a delicate constitution is very similar to that of the emetic. Both treatments would be appropriate for the Buddha, who, as a Buddhist ascetic accustomed to eating light foods, was one with a mild constitution.[31] Whether employed as an emetic or as a purgative, the therapeutic technique demonstrates a continuity between early Buddhist medicine and early *āyurveda*.

The principal difference between the āyurvedic presentation of this treatment and the Buddhist account of Jīvaka's healing centers on the technical terminology. In the Pāli text, the procedure is called *virecana*, while the medical authors' term is *vamana*. Clearly, it is an emetic (*vamana*). Caraka, in his definition of the two evacuative techniques, offers information that may help to explain why this is so: "Thereupon that which removes the humors from the upper part [of the body] is called emetic [*vamana*]; that which [removes the humors] from the lower part [of the body] is called purgative [*virecana*]; or both receive the name *virecana* because they remove [lit. purge] impurities from the body."[32] The use of *virecana* in the Pāli text reflects precisely the generic meaning of *virecana* in Caraka and hence points to the close similarity between evacuation therapy in the early āyurvedic tradition and in the Buddhist account of the lay Buddhist physician Jīvaka Komārabhacca.

The episode of Jīvaka's healing the Buddha also occurs in the *Mūlasarvāstivāda Vinaya*, preserved in Sanskrit and Tibetan, and in the Sarvāstivāda and Mahīśāsaka Vinayas, preserved in Chinese. In the Sanskrit-Tibetan version, the Buddha, the King of the Himavant Mountains, because of his constant contact with snow, suffered chills and developed an illness (*glāna*) with oozing (i.e., runny nose?) (*abhiṣyanda*). Jīvaka made him smell thirty-two lotuses (*utpala*) infused with purgative drugs (*sraṃsaṇīya dravya*). This resulted in thirty-two evacuations and a complete purgation of the "peccant" humors (*doṣa*s), that is, those that are loosened (i.e., moved to another seat) (*cyuta*) but not flowing (i.e., liquefied or dissolved) (*na sruta*), those that are flowing but not loosened, those that are both loosened and flowing, and those that are neither loosened nor flowing. Jīvaka then instructed the Buddha to eat only yellow myrobalan (*harītakī*) with treacle (*guḍa*) and to have a regular regimen of cream (*maṇḍa*). The Buddha did this and was restored to a healthy condition.[33]

According to the Chinese versions, the Buddha suffered from cold sweats[34] and required a purgative. Ānanda was sent to fetch Jīvaka, who, when informed of the Buddha's illness, thought to himself: "The Buddha's virtue is great, and it would not do to use herbs [i.e. "tree medicine"] or pain medicine as would be the rule for others. I must take a special medicinal dust which falls from the blue lotus flower and give it to the Buddha."[35] After preparing the medicine, Jīvaka approached the Buddha with the purgative made of the lotus (*uppala* [*utpala*]) dust and instructed him to inhale it three times, causing thirty purgations.[36] The Buddha inhaled the purgative and evacuated twenty-nine times. Jīvaka told him to drink warm water, resulting in the thirtieth purge. Jīvaka then prepared a medicinal food and drink consisting of soft rice, gruel, and broth, and fed a specific quantity of it to the Buddha, who, after taking the food, recovered completely and regained his strength and normal complexion.[37]

These versions follow fairly closely that found in the Pāli Vinaya. The Sanskrit-Tibetan account, however, is more explicit about the nature of the Buddha's illness. The oozing (*abhiṣyanda*) symptom corresponds to the variant reading of the affliction of the "peccant" humors found in the *Caraka Saṃhitā*: *doṣair abhiṣyannas* yo *naraḥ* (a man who is oozing [or moistened] with the "peccant" humors).[38] The fourfold classification of the "peccant" humors indicates a well-developed etiological doctrine based on the humoral theory. The Chinese versions stress the Buddha's unique physical characteristics, emphasizing that he is not like an ordinary human and therefore requires special treatment. Likewise, the Chinese accounts indicate an imperfect knowledge of the principles and practices of Indian medicine.

Among its many virtues, Suśruta mentions that exercise (*vyāyāma*) promotes digestion even of disagreeable food and of digested and undigested food without exciting the humors, and that it is always suitable in cases of healthy people who have consumed greasy food (*snigdhabhojana*), especially in spring and summer.[39] Both cases involving the "peccant" humors treated by Jīvaka find correspondences in the early medical treatises, indicating a common source of medical practice.

APPENDIX II

Glossary of Pāli and Sanskrit Plant Names

This glossary contains a list of plant names encountered in the various Pāli and Sanskrit sources utilized in this study. They are arranged in Sanskrit alphabetical order, and each entry includes both its common name and Linnaean nomenclature. Pāli terms are provided with their Sanskrit equivalents, which are cross-referenced to the Pāli entry. A note of caution, however, applies to this and other glossaries of plant names found in ancient Indian literature. Because botany in ancient India was not an exact science, modern equivalents for the antique Pāli and Sanskrit terms are only approximations based on the examination of more recent Indian pharmacopoeias (*nighaṇṭus*) and the work of botanists knowledgeable in both Indian materia medica and the modern science of plant taxonomy. Scientists and Sanskritists are only now beginning to work together to obtain a better identification of plants. The results of their combined efforts will provide more definitive knowledge of ancient Indian botany.

agnimantha   headache tree; *Premna latifolia* Roxb., var. *mucronata* Clarke.

ativisa (Skt. ativiṣā)   Indian atees; *Aconitum heterophyllum* Wall., or *A. palmatum* D. Don.

ativiṣā   s.v. ativisa.

abhayā   s.v. harītaka.

aśvatthá   sacred fig tree; *Ficus religiosa* Linn.

āmalaka (Skt. āmalakī)   emblic myrobalan, Indian gooseberry; *Phyllanthus emblica* Linn.

āmalakī   s.v. āmalaka.

āragvadha   drumstick or purging cassia; *Cassia fistula* Linn.

ārdraka   s.v. siṅgivera.

utpala   blue lotus; *Nymphaea stellata* Willd.

udumbara cluster fig; *Ficus glomerata* Roxb.

uśīra s.v. usīra.

usīra (Skt. uśīra) vetiver; *Vetiveria zizanioides* (Linn.) Nash = *Andropogon muricatus* Retz.

eraṇḍa castor; *Ricinus communis* Linn.

kaṭuka s.v. kaṭukarohiṇi.

kaṭukarohiṇi (Skt. kaṭukarohiṇī = kaṭuka) black hellebore; *Picrorrhiza kurroa* Royle ex Benth.

kaṭukarohiṇī s.v. kaṭukarohiṇi.

kaṇḍala ? white mangrove; *Avicennia officinalis* Linn.

kappāsikā (Skt. kărpāsikā or kārpāsī) cotton tree; *Gossypium herbaceum* Linn.

karañja s.v. nattamāla.

kalāya wild pea; *Lathryus sativus* Linn.

kărpāsikā s.v. kappāsikā.

kārpāsī s.v. kappāsikā.

kāśmari white teak; *Gmelina arborea* Linn.

kiṃśuka s.v. kiṃsuka.

kiṃsuka (Skt. kiṃśuka) = palāśa dhak tree; *Butea monosperma* (Lam.) Taub. = *B. frondosa* Roxb.

kuṭaja kurchi tree; *Holarrhena antidysenterica* Wall.

kulattha horse gram; *Dolichos biflorus* Linn.

kuśa kuśa grass; *Desmostachya bipinnata* Stapf, or *Poa cynosuroides* Retz.

kuṣṭha costus; *Saussurea lappa* C. B. Clarke.

kusumbha safflower; *Carthamus tinctorius* Linn.

kṛtavedhana ribbed luffa, ridged gourd; *Luffa acutangula* (Linn.) Roxb.

kṛṣṇasārivā black creeper; *Cryptolepis buchanani* Roem. et Schutt.

kṛṣṇā s.v. pippala.

ketaka fragrant screwpine; *Pandanus odoratissimus* Roxb.

kodrava kodra; *Paspalum scrobiculatum* Linn.

khajjūtĭ (= ? Skt. kharjūrĭ) wild date; *Phoenix sylvestris* Roxb.

khadira catechu tree; *Acacia catechu* Willd. = *Mimosa catechu* Linn.

candana sandalwood; *Santalum album* Linn.

citraka leadwort; *Plumbago zeylanica* Linn.

jambura blackberry; *Eugenia jambolana* Lam.

ṭiṇṭuka = ṭuṇṭŭka = śyonāka Indian calosanthes; *Oroxylum indicum* Vent.

ṭuṇṭŭka s.v. ṭiṇṭuka.

tagara ? East India rosebay; *Ervatamia divaricata* Burkill = *Tabernaemontana coronaria* Willd.

taṇḍula rice; *Oryza sativa* Linn.

tālīsa silver fir; *Abies webbiana* Lindl.

tila sesame; *Sesamum indicum* Linn.

tilvaka s.v. lodda.

tuṅgahara ? perhaps a type of thorny tree with yellow exudation.

tulasī s.v. sulasī.

triphalā three myrobalans, i.e., harītakī, āmalaka, tilvaka (s.v.).

trivṛt (red-rooted) turpeth tree; *Operulina turpethum* (Linn.) Silva Manso.

dantinī s.v. dantī.

dantī = dantinī wild croton; *Croton tiglum* Linn.

dāruharidrā Indian barberry; *Berberis aristata* DC.

dhāmārgava large acute-angled cucumber; *Luffa cylindrica* (Linn.) M. Roem.

naktamāla s.v. nattamāla.

nattamāla (Skt. naktamāla = karañja) Indian beech; *Pongamia pinnata* (Linn.) Merr.

nāḍīhiṅgu s.v. hiṅgusipāṭikā.

nikumbhā turpeth tree; *Baliospermum montanum* Meul.-Arg.

nimba Indian lilac, neem; *Azadirachta indica* A. Juss. = *Melia azadirachta* Linn.

pakkava = ? latā s.v. priyaṅgu.

paṭola wild snake gourd; *Trichosanthes dioica* Roxb., or *T. cucumerina* Linn.

pathyā s.v. harītaka.

palāśa s.v. kiṃsuka.

pāṭalā trumpet flower tree; *Stereospermum suaveolens* DC.

pippala (var. pippali) (Skt. pippalī) = kṛṣṇā long pepper; *Piper longum* Linn.

pippalī s.v. pippala.

priyaṅgu = latā (Ḍalhaṇa to SuUtt 60.48) perfumed cherry; *Callicarpa macrophylla* Vahl.; or a type of grass, *Setaria italica* Beauv.

balvajā̆ balvaja grass; *Pollinidium angustifolium* Comb. Nov.

bilva Bengal quince; *Aegle marmelos* Corr.

bhaddamuttaka (Skt. bhadramusta = mustaka) nutgrass; *Cyperuss rotundus* Linn. = *C. scariosus* R.Br.

bhadramusta s.v. bhuddamuttaka

bhūnimba s.v. yavatiktā.

madana emetic nut; *Randia dumetorum* Linn.

madhuka (Skt. madhūka) mahua; *Bassia longifolia* Linn., or *B. latifolia* Roxb.

madhūka s.v. madhuka.

marica black pepper; *Piper nigrum* Linn.

māṣa s.v. māsa.

māsa (Skt. māṣa) black gram; *Phaseolus mungo* Linn.

mugga (Skt. mudga) green gram; *Phaseolus aureus* Roxb.

muñja muñja grass; *Saccharum munja* Roxb.

mudga s.v. mugga.

mustaka s.v. bhaddamuttaka.

mūrvā bowstring hemp; *Sansevieria roxburghiana* Schult. f. = *S. zeylanica* Roxb.

meṣaśṛṅgī "ram's horn"; *Dolichandrone falcata* Seem.

yavatiktā = ? bhūnimba kalmegha; *Andrograhis paniculata* Nees (cf. śaṅkhinī).

rajanī s.v. haliddā.

rālā̆ ral tree; *Mimosa rubicaulis* Linn.

latā s.v. priyaṅgu.

lodda (Skt. lodhra) = tilvaka lodh tree; *Symplocos racemosa* Roxb.

lodhra s.v. lodda.

vaca (Skt. vacā) sweet flag, orrisroot; *Acorus calamus* Linn.

vacattha (var. vacatta) = ? sesavaca (var. setavaca) (Skt. śvetavacā) ? white sweet flag; perhaps a variety of *Acorus calamus* Linn.

vacā = śvetavacā = śveta = haimavatī (Ḍalhaṇa) s.v. vaca.

vaṃśa bamboo; *Bambusa arundinacea* Willd.

viḍaṅga s.v. vilaṅga.

vibhītaka (Skt. v[b]ibhītaka) beleric myrobalan; *Terminalia bellerica* Roxb.

v(b)ibhītaka s.v. vibhītaka.

vilaṅga (Skt. viḍaṅga) embelia; *Embelia ribes* Burm., var. *E. robusta* Roxb.

śaṅkhinī canscora; *Euphorbia dracunculoides* Lam.; or = a type of yavatiktā (Ḍalhaṇa) (s.v.).

śaṇa śaṇa hemp, Bombay hemp; *Crotalaria juncea* Linn.

śārṅgaṣṭā jequirity; *Abrus prectatorius* Linn.; or a species of karañja (s.v.); or *Dregea volubilis* Benth (see Sharma, *Ḍalhaṇa and His Comments on Drugs*, 174).

śāla sal tree; *Shorea robusta* Gaertn.

śṛṅgavera s.v. siṅgivera.

śyāma black turpeth; a variety of trivṛt (s.v.).

śyonāka s.v. ṭiṇṭuka.

śveta s.v. vacā, vacattha.

śvetavaca s.v. vacā, vacattha.

saptalā golden cleome; *Euphorbia pilosa* Linn.; or = a type of sehuṇḍa (snuhī) (s.v.).

sarja Indian copal tree; *Vateria indica* Linn.

sarṣapa s.v. sāsapa.

sāsapa (Skt. sarṣapa) mustard; *Bissia campestris* Linn, var. *sarson* Prain.

sughā s.v. sehuṇḍa.

surasī(ā) s.v. sulasī.

sulasī (var. sulasā) (Skt. surasī[ā] = tulasī) holy basil; *Ocimum sanctum* Linn.

siṅgivera (Skt. śṛṅgavera = ārdraka) ginger; *Zingiber officinale* Roscoe.

sehuṇḍa = snuhī = sughā common milk hedge; *Euphorbia neriifolia* Linn (cf. saptalā).

setavaca s.v. vacā, vacattha.

sesavaca s.v. vacā, vacattha.

snuhī s.v. sehuṇḍa.

harītaka (Skt. harītakī = pathyā = abhayā) cherbulic or yellow myrobalan, Indian gall nut; *Terminalia chebula* Retz.

harītakī s.v. harītaka.

haridrā s.v. haliddā.

haliddā (Skt. haridrā = rajanī) turmeric; *Curcuma longa* Linn.

hiṅgu asafetida; *Ferula foetida* Regel. = *F. asafoetida* Linn.

hiṅguparṇī s.v. hiṅgusipāṭikā.

hiṅguśivāṭika s.v. hiṅgusipāṭikā.

hiṅgusipāṭikā (? Skt. hiṅguśivāṭika = hiṅguparṇī = nāḍīhiṅgu) *Gardenia gummifer* Linn.

haimavatī s.v. vacā.

Notes

Abbreviations

AB	*Aitareya Brāhmaṇa*
AH	*Aṣṭāṅgahṛdaya Saṃhitā*
AN	*Aṅguttaranikāya*
Ap	*Apadāna*
AS	*Aṣṭāṅgasaṃgraha*
AV	*Atharvaveda* (Śaunaka recension)
BD	*The Book of the Discipline* (Vinaya Piṭaka)
Bh	*Bhelasaṃhitā*
BKS	*The Book of Kindred Sayings* (*Saṃyuttanikāya*)
Ca	*Caraka Saṃhitā*
Ci	*Cikitsāsthāna*
CV	*Cullavagga*
DB	*Dialogues of the Buddha* (*Dīghanikāya*)
DhNi	*Dhanvantarinighaṇṭu*
DN	*Dīghanikāya*
GS	*The Book of Gradual Saying* (*Aṅguttaranikāya*)
In	*Indriyasthāna*
Jā	*Jātaka*
Ka	*Kalpasthāna*
KaiNi	*Kaiyadevanighaṇṭu*
Kaps	*Kapiṣṭhala Kaṭha Saṃhitā*
KB	*Kauśītaki Brāhmaṇa*
Khp	*Khuddaka-pāṭha*
KS	*Kāṭhaka Saṃhitā*

Miln	*Milindapañha*
MLS	*Middle Length Sayings* (*Majjhimanikāya*)
MN	*Majjhimanikāya*
MQ	*Milinda's Questions* (*Milindapañha*)
MS	*Maitrāyaṇī Saṃhitā*
MV	*Mahāvagga*
Nadk	Nadkarni's *Indian Materia Medica*, 2 vols.
Ni	*Nidānasthāna*
Nidd I	*Mahāniddesa*
Nidd II	*Cullaniddesa*
PED	*The Pāli Text Society's Pāli–English Dictionary*
Ps	Buddhaghosa's *Papañcasūdāni*
PTC	*Pāli Tripiṭakam Concordance*
RV	*Ṛgveda*
Śā	*Śārīrasthāna*
ŚB	*Śatapatha Brāhmaṇa*
Si	*Siddhisthāna*
SN	*Saṃyuttanikāya*
Sn	*Suttanipāta*
SoNi	*Soḍhalanighaṇṭu*
Spk	*Sāratthappakāsinī*
Su	*Suśruta Saṃhitā*
Sū	*Sūtrasthāna*
Sv	Buddhaghosa's *Sumaṅgalavilāsinī*
T	*Taisho Tripiṭaka* (The Chinese Tripiṭaka)
Tha	*Theragāthā*
Thī	*Therīgāthā*
Toḍ	Toḍarānanda's *Āyurveda Saukhyaṃ*
TS	*Taittirīya Saṃhitā*
Utt	*Uttarasthāna, Uttaratantra*
VA	Buddhaghosa's *Samantapāsādikā*
Vi	*Vimānasthāna*

Introduction

1. Several modern scholars have importantly enlarged our understanding of the history of Indian physical sciences. David Pingree's studies show that the exact sciences in ancient and medieval India were exclusively part of the brāhmaṇic domain yet adapted much information from ancient Western scientific traditions; see especially his "Astronomy and Astrology in India and Iran," *Isis*

54 (1963): 229–46; "The Mesopotamian Origin of Early Indian Mathematical Astronomy," *Journal for the History of Astronomy* 4 (1973): 1–12; "The Recovery of Early Greek Astronomy from India," *Journal for the History of Astronomy* 7 (1976): 109–23; "History of Mathematical Astronomy in India," in Charles C. Gillispie, ed., *Dictionary of Scientific Biography* (New York: Scribner, 1978), 15: 533–633; and his edition and translation of *The Yavanajātaka of Sphujidhvaja*, 2 vols. (Cambridge, Mass.: Harvard University Press, 1978). The many books and essays of J. Frits Staal present a unique understanding of Indian science as closely connected to brāhmaṇic intellectual traditions of ritual and language; see especially his *Science of Ritual* (Puṇe: Bhandarkar Oriental Research Institute, 1982); *The Fidelity of Oral Tradition and the Origins of Science* (Amsterdam: North Holland, 1986); and *Universals: Studies in Indian Logic and Linguistics* (Chicago and London: University of Chicago Press, 1988). Debiprasad Chattopadhyaya's *History of Science and Technology in Ancient India. The Beginnings* (Calcutta: Firma KLM, 1986) demonstrates the close association between the development of mathematics and astronomy and the Vedic sacrificial cults, although his overspeculative and highly theoretical effort to correlate periods of ancient India urbanization with successive stages of Indian scientific development is unreliably based on inconclusive data. A useful overview is D. M. Bose, ed., *A Concise History of Science in India* (New Delhi: Indian National Science Academy, 1971).

2. Jean Filliozat, *La Doctrine classique de la médecine indienne, ses origines et ses parallèles grecs* (Paris: Imprimerie Nationale, 1949), 2–20 (English, 2–25). Sheldon Pollock demonstrates that the brāhmaṇic technique of establishing divine origin is characteristic of most of the Hindu didactic or śāstric literature ("The Theory of Practice and the Practice of Theory in Indian Intellectual History," *Journal of the American Oriental Society* 105 [1985]: 499–519).
3. Thomas S. Kuhn, *The Structure of Scientific Revolutions* (Chicago: University of Chicago Press, 1962).
4. In a recent article, Richard Gombrich persuasively argues that the Buddhist monastery had the required organizational structure from its inception to carry out the function of systematizing and preserving Buddhist scripture. This structure then facilitated the codification and transmission of the Indian medical knowledge as part of Buddhist religious literature ("How the Mahāyāna Began," *Journal of Pāli and Buddhist Studies* 1 [1988]: 29–46).
5. Debiprasad Chattopadhyaya, *Science and Society in Ancient India* (Calcutta: Research India Publications, 1977).
6. Jyotir Mitra, *A Critical Appraisal of Āyurvedic Material in Buddhist Literature, with Special Reference to Tripiṭaka* (Varanasi: The Jyotiralok Prakashan, 1985).
7. Important to any comparative studies between Indian, Hellenistic, and Chinese medical systems are Francis Zimmermann's recent analysis of similarities and differences between Indian and Hellenistic medical epistemologies (*The Jungle and the Aroma of Meats* [Berkeley: University of California Press, 1987], 31–33,

97, 129–33, 196–98 [French, 44–46, 114, 145–49, 216–18]), and Nathan Sivin's useful survey of past and current trends in the study of ancient Chinese science and medicine, which indicates possible avenues for comparative studies of Chinese and Indian medicine and provides an excellent bibliography for such endeavors ("Science and Medicine in Imperial China—The State of the Field," *The Journal of Asian Studies* 47 [1988]: 41–90).

Chapter 1

1. The principal sources on Indus Valley archaeology are the following: Sir John Marshall, ed., *Mohenjo-Dāro and the Indus Civilization*, 3 vols. (London: Arthur Probsthain, 1931); Ernest J. H. Mackay, *Further Excavations at Mohenjo-Dāro*, 2 vols. (Delhi: Manager of Publications, Government of India Press, 1943); E. J. H. Mackay, *Chanhu-Dāro Excavations, 1935–36* (New Haven, Conn.: American Oriental Society, 1943); Madho Sarup Vats, *Excavations at Harappā*, 2 vols. (Delhi: Manager of Publications, Government of India Press, 1940); S. R. Rao, *Lothal and the Indus Civilization* (New York: Asia Publishing House, 1973); T. G. Avaramuthan, *Some Survivals of the Harappan Culture* (Bombay: Karnatak Publishing House, 1942); Sir Mortimer Wheeler, *Early India and Pakistan to Aśoka*, rev. ed. (New York: Praeger, 1968); Bridget Allchin and Raymond Allchin, *The Rise of Civilization in India and Pakistan* (London: Cambridge University Press, 1985); George F. Dales, "The Decline of the Harappans," in C. C. Lamberg-Karlovsky, ed., *Old World Archaeology: Foundations of Civilizations; Readings from Scientific American* (San Francisco: Freeman, 1972), 157–64; Walter A. Fairservis, Jr., *The Roots of Ancient India*, 2d rev. ed. (Chicago: University of Chicago Press, 1975); Doris Srinivasan, "The So-called Proto-Śiva Seal from Mohenjo-Dāro: An Iconographic Assessment," *Archives of Asian Art* 29 (1975–76): 47–58; Gregory L. Possehl, ed., *Ancient Cities of the Indus* (Durham, N.C.: Carolina Academic Press, 1979); and D. P. Agrawal, *The Archaeology of India* (London and Malmö: Curzon Press, 1985).
2. At Kālibangan, a northern Indus valley site, evidence indicates that rituals involving fire were performed (see Allchin and Allchin, *Rise of Civilization in India and Pakistan*, 216–17; Agrawal, *Archaeology of India*, 156).
3. One such figure with three faces (one confronting the viewer and two in profile) and surrounded by different animals prompted Sir John Marshall, the excavator of Mohenjo-dāro, to speculate that it represented an early form of the god Śiva, later known as the "Lord of Beasts" (*paśupati*) (*Mohenjo-Dāro and the Indus Civilization*, 1: 52–56). Although soundly reasoned, Marshall's suggestion is untenable: Śiva was not a deity incorporated into the Vedic religion from the outside but evolved from Rudra, a god from the mainstream of the Vedic pantheon (see, in particular, Doris M. Srinivasan, "Vedic Rudra-Śiva," *Journal of the American Oriental Society* 103 [1983]: 543–56,

and "Unhinging Śiva from the Indus Civilization," *Journal of the Royal Asiatic Society of Great Britain and Ireland* [1984]: 77–89).

4. E. T. Kirby defines the shaman as "a 'master of spirits' who performs in trance, primarily for the purpose of curing the sick by ritualistic means" (*Ur-drama: The Origins of Theatre* [New York: New York University Press, 1975], 1); see also Mircea Eliade, *Shamanism: Archaic Techniques of Ecstasy*, trans. Willard R. Trask (Princeton, N.J.: Princeton University Press, 1972), 4 and passim.
5. See in particular, Herman Grapow, *Grundriss der Medizin der alten Ägypter*, vols. 1–4 (Berlin: Akademie Verlag, 1954, 1956, 1958, 1959); B. Ebbell, trans., *The Papyrus Ebers* (Copenhagen: Levin & Munksgaard, 1937); George Contenau, *La Médecine en Assyrie et en Babylone* (Paris: Libraire Maloine, 1938); René Labat, ed. and trans., *Traité akkadien de diagnostics et pronostics médicaux*, vols. 1, 2 (Paris: Academie Internationale d'Histoire des Sciences; Leiden: Brill, 1951); and Henry E. Sigerist, *A History of Medicine: Primitive and Archaic Medicine* (New York: Oxford University Press, 1955), 1: 217–497.
6. See K. G. Zysk, *Religious Healing in the Veda* (Philadelphia: American Philosophical Society, 1985), 3–4.
7. The principal editions of the *Ṛgveda* include F. Max Müller, ed., *The Hymns of the Rig-Veda with Sāyaṇa's Commentary*, 2d ed., 4 vols. (1890–92; reprint, Varanasi: The Chowkhamba Sanskrit Series Office, 1966); and, in romanized script, Theodor Aufrecht, ed., *Die Hymnen des Rigveda*, 2 vols. (1887; reprint, Wiesbaden: Otto Harrassowitz, 1968). The best complete translation is Karl F. Geldner, *Der Rig-Veda*, 3 pts. (Cambridge, Mass.: Harvard University Press, 1951). An inferior but complete English rendering is Ralph T. H. Griffith, trans., *The Hymns of the Ṛgveda*, 5th ed., 2 vols. (Varanasi: The Chowkhamba Sanskrit Series Office, 1971). Questions pertaining to the Āryan invasions of the Indian subcontinent and to the original homeland of the Indo-Europeans are constantly undergoing critical reassessment (see, in particular, Colin Renfrew, *Archaeology and Language: The Puzzle of Indo-European Origins* [Harmondsworth: Penguin Books, 1989], esp. 178–210).
8. RV 10.97.
9. The text of the extant *Atharvaveda* occurs in two forms: the *Śaunaka*, which has been edited with Sāyaṇa's fourteenth-century Sanskrit commentary and completely translated into English and partially into other European languages, and the *Paippalāda*, which is preserved in an imperfectly edited Kaśmīr recension and in a partially edited Orissā recension. Neither of the *Paippalāda* texts has been translated. The principal editions of the *Śaunaka* recension include Vishva Bandhu et al., eds., *Atharvaveda (Śaunaka) with the Pada-pāṭha and Sāyaṇācarya's Commentary*, 4 vols. (Hoshiarpur: Vishvesvaranand Vedic Research Institute, 1960–62); and R. Roth and W. D. Whitney, eds., *Atharva-Veda Saṃhitā* (Berlin: Ferd. Dümmlers Verlagsbuchhandlung, 1924). The best English translation is W. D. Whitney, trans., Charles R. Lanman, ed., *Atharva-veda-saṃhitā*, 2 pts. (1905; reprint, Delhi: Motilal

Banarsidass, 1971). Selections are rendered in English by Maurice Bloomfield, *Hymns of the Atharvaveda* (1897; reprint, Delhi: Motilal Banarsidass, 1964). Books 1–5 have been translated into German by Albrecht Weber (*Indische Studien*, 4 [1858]: 393–430, 13 [1873]: 129–216, 17 [1885]: 177–314, 18 [1898]: 1–288); books 7–13, into French by Victor Henry (Paris: J. Maisonneuve, 1891–96). Selections have been rendered into Russian by T. Ja. Elizarenkova, *Atxarvaveda* (Moscow: Nauka, 1976); and into Spanish by Fernando Tola, *Himno del Atharvaveda* (Buenos Aires: Editorial Sudamericana, 1968). The Kaśmīr recension of the *Paippalāda* was edited by LeRoy Carr Barrett, "The Kashmirian Atharva Veda," books 1–5, 7–15, 18, in the *Journal of the American Oriental Society* 26 (1905): 197–295, 30 (1909–10): 187–258, 32 (1912): 343–90, 35 (1915): 42–101, 37 (1917): 257–308, 40 (1920): 145–69, 41 (1921): 264–89, 42 (1922): 105–46, 43 (1923): 96–115, 44 (1929): 258–69; 46 (1926): 34–48, 47 (1927): 238–49, 48 (1928): 34–65, 50 (1930): 43–73, 58 (1938): 571–614; books 16–17 and 19–20 as separate monographs (New Haven, Conn.: American Oriental Society, 1936, 1941); and book 6 was edited by Franklin Edgerton, *Journal of the American Oriental Society* 34 (1915): 374–411. It was compiled and edited in *devanāgarī* characters by Raghu Vira, *Atharvaveda of the Paippalāda* (1936–42; reprint, Delhi: Arsh Sahitya Prashar Trust, 1979). Only the first four books on the Orissā recension of the *Paipalāda* have appeared: Durgamohan Bhattacharyya, ed., *Atharvavedīyā Paippalāda Saṃhitā, Kāṇḍas* 1–4 (Calcutta: Sanskrit College, 1964, 1970). The remainder of the Orissā Manuscripts of the *Paippalāda* recension are currently being edited and should appear in due course.

10. Maurice Bloomfield, ed., *The Kauśika Sūtra of Altharvaveda* (1889; reprint, Delhi: Motilal Banarsidass, 1972). Partial translations by Willem Caland, *Altindisches Zauberritual* (1900; reprint, Wiesbaden: Martin Sändig, 1967); Bloomfield, *Hymns of the Atharvaveda*, 233–692; and J. Gonda, *The Savayajñas* (*Kauśikasūtra* 60–68) (Amsterdam: N. V. Noord-Hollandsche Uitgevers Maatschappij, 1965).
11. Zysk, *Religious 'Healing in the Veda*, 1–11.
12. See, in particular, Arrian, *Indica* (15.11–12), translated and discussed in K. G. Zysk, "The Evolution of Anatomical Knowledge in Ancient India, with Special Reference to Cross-cultural Influences," *Journal of the American Oriental Society*, 106 (1986): 695. The entire subject of ancient Indian toxicology requires a detailed examination, tracing its development from the Veda to *āyurveda*.
13. See K. G. Zysk, "Towards the Notion of Health in the Vedic Phase of Indian Medicine," *Zeitschrift der Deutschen Morgenländischen Gesellschaft* 135 (1985): 312–18.
14. See Zysk, "Evolution of Anatomical Knowledge in Ancient India," 687–705.
15. See Mircea Eliade, *Le Sacré et le profane* (Paris: Gallimard, 1965), 60–98, and *The Myth of the Eternal Return*, trans. Willard R. Trask (Princeton, N.J.: Princeton University Press, 1971), passim; cf. his *Quest: History and Meaning in Religion* (Chicago: University of Chicago Press, 1969), 72–87, and *Traité d'histoire des religions* (Paris: Petit Bibliothèque Payot, 1975), 326–66.

16. Zysk, *Religious Healing in the Veda*, 73.
17. U. C. Dutt and George King, *The Materia Medica of the Hindus*, rev. ed. (Calcutta: Mandan Gopal Dass, 1922), 277–78.
18. Ibid., 181–82.
19. See Zysk, *Religious Healing in the Veda*, 40, 72–74, 97–98, and passim.
20. See A. A. Macdonell, *Vedic Mythology* (1898; reprint, Delhi: Motilal Banarsidass, 1974), 104–15.
21. See Zysk, *Religious Healing in the Veda*, 39–40.
22. See, in particular, RV 10.97 and AV 8.7, the contents of which imply an intimate knowledge of local plant life and the worship of vegetation in the form of a goddess or goddesses. See also Zysk, *Religious Healing in the Veda*, 99–102, 238–56.
23. There is RV 10.97, which is a very late hymn, characteristic of Atharvavedic hymns. The divine Soma plant was recognized principally for its religious value and for the role it played in Vedic ritual. It is mentioned as a medicine only four times (RV 8.72.17, 79.2, and in the very late hymns of RV 10.25.11, 10.97.18).
24. See K. G. Zysk, "Mantra in *Āyurveda*: A Study of the Use of Magico-Religious Speech in Ancient Indian Medicine," in Harvey Alper, ed., *Mantra* (Albany: State University of New York Press, 1989), 123–43.
25. See Chapter 6, on Nonhuman Disease, 87–88.
26. See Chapter 4, on Medical Knowledge in Nonmonastic Buddhist Treatises, 61–73.

Chapter 2

1. RV 9.112.1.
2. RV 10.97.6, 22.
3. Chattopadhyaya, *Science and Society in Ancient India*, 235.
4. The Bahiṣpavamāna is the name of a Stoma or Stotra sung at the Soma sacrifice for purification of the sacrificers before entering the grounds of the sacred Sadas. See Julius Eggeling, trans., *The Śatapatha Brāhmaṇa*, pt. 2 (1885; reprint, Delhi: Motilal Banarsidass, 1972), 310 n.; cf. TS 3.1.10; AB 3.1 (= 11.1); KB 8.7; cf. AB 3.14 (= 11.3). See also A. B. Keith, trans., *The Veda of the Black Yajus School Entitled Taittirīya Sanhitā*, pt. 1 (1914; reprint, Delhi: Motilal Banarsidass, 1967), cviii, cxv. Here, however, it appears to be used particularly for the purification of physicians.
5. TS 6.4.9.1–3. Variants are found at KapS 42.5; KS 27.4; MS 4.6.2; and ŚB 4.1.5.13–16; 14.1.1.13–26. In none of these is it specifically stated that medicine is not to be practiced by a Brāhmaṇ. This proscription is particular to the TS. The KapS and KS follow each other very closely and mention that the Aśvins placed that which is their healing form (*tanu*) in the three places. The MS, which varies most greatly from the others, simply states that medicine

was put in the three places, without specifying by whom. The KapS, KS, and MS add that whoever is unfriendly must be excluded from the Bahiṣpavamāna. The entire ritual prescription and exegesis must refer to the myth found in the *Ṛgveda*, which mentions how the Aśvins were given a drink of Soma (i.e., *madhu*) when they gave a horse's head to the *átharvan's* son Dadhyañc (RV 1.116.12, 117.22, 119.9, etc.). On the entire mythic story, see M. Witzel, "On the Origin of the Literary Device of the 'Frame Story' in Old Indian Literature," in Harry Falk, ed., *Hinduismus und Buddhismus* (Freiburg: Hedwig Falk, 1987), 380–414.

6. KapS 4.25, KS 27.4, and MS 4.62. The variants found in the KapS and KS state that the waters should be enchanted with the following charm: "Whichever sick [man] one desires, [him] one enlivens" (*yaṃ kāmayet āmayāvinaṃ jīved*). Cf. RV 10.125.5 (= AV 4.30.30), where the goddess of speech (Vāc) says: "Whomever I desire, him I make best, him a priest, him a seer, him very wise" (*yáṃ kāmáye táṃtam ugráṃ kṛṇomi táṃ brahmā́ṇaṃ táṃ ṛ́ṣiṃ táṃ sumedhā́m*). Cf. also TS 1.7.1.3–4.
7. ŚB 4.1.5.14.
8. *Manusmṛti* 3.108, 152; 4.212, 220. See also Chattopadhyaya, who discusses the material from a slightly different angle (*Science and Society in Ancient India*, 212–51).
9. Bloomfield, *Hymns of the Atharvaveda*, xxxix–xl.
10. Filliozat, *La Doctrine classique de la médecine indienne*, 15–17 (English, 19–21).
11. Chattopadhyaya, *Science and Society in Ancient India*, 270–306.
12. Ibid., 29.
13. See CaVi 8.13, 20.
14. See CaSū 1.1–40 and SuSū 1.1–6, 17, 41. See also Pollock, "Theory of Practice and the Practice of Theory in Indian Intellectual History," 513.
15. The version recounted here refers to the sacrifice of Dakṣa found in the *Mahābhārata* and the *Varuṇa Purāṇa*. The Epic and Purāṇic accounts, however, do not mention the Aśvins, suggesting that the author of the medical passage had both the Vedic and the later Hindu versions in mind (see A. F. R. Hoernle, trans., *The Suśruta Saṃhitā*, fasc. 1 [Calcutta: Asiatic Society, 1897], 5–6 n. 13; and G. D. Singhal et al., trans., *Fundamentals and Plastic Surgery Considerations in Ancient Indian Surgery* [Varanasi: Singhal Publications, 1981], 25 n. 2).
16. SuSū 1.17–20. The translation follows the commentary of Ḍalhaṇa, who glosses Prajāpati as Dakṣa.
17. CaSū 30.21; cf. SuSū 1.6.
18. Chattopadhyaya, *Science and Society in Ancient India*, 2, 18, 35–36, 40–44, 365–424.
19. See R. H. Robinson and W. L. Johnson, *The Buddhist Religion: A Historical Introduction*, 3d ed. (Belmont, Calif.: Wadsworth, 1982), 7; A. L. Basham, "The Background to the Rise of Buddhism," in A. K. Narain, ed., *Studies in History of Buddhism* (Delhi: B. R. Publishing, 1980), 13–17; A. K. Warder, *Indian*

Buddhism (Delhi: Motilal Banarsidass, 1970), 33–36; and G. C. Pande, *Studies in the Origins of Buddhism* (Allahabad: University of Allahabad, 1957). See also E. J. Thomas, *The History of Buddhist Thought* (London: Routledge & Kegan Paul, 1959), 77 ff., and L. de la Vallée-Poussin, *Histoire du monde*, Vol. 3: *Indo-européens et Indo-iraniens; L'Inde jusque vers 300 av. J.-C.* (Paris: Editions de Boccard, 1924), 301, 304–14.

20. Warder, *Indian Buddhism*, 33–35; See also A. K. Warder, "On the Relationship Between Early Buddhism and Other Contemporary Systems," *Bulletin of the School of Oriental and African Studies* 17 (1956): 43–63.
21. Heinz Bechert presents compelling evidence for placing the death of the Buddha in the second quarter of the fourth century B.C.E. ("The Date of the Buddha Reconsidered," *Indologica Taurinensia* 10 [1982]: 29–26). See also the forthcoming article by Gananath Obeyesekere, "Myth, History and Numerology in the Buddhist Chronicles"; see also Heinz Bechert, "A Remark on the Problem of the Data of Mahāvīra," *Indologica Taurinensia* 11 (1983): 287–90.
22. *Bhagavatī Sūtra* 15.539, 658–69.
23. A. L. Basham, *History and Doctrines of the Ājīvikas* (London: Luzac, 1951), 56–58. See also Warder, "On the Relationship Between Early Buddhism and Other Contemporary Systems," 51.
24. DN 1.1.27(1, 12): *...vassakammaṃ vossakammaṃ...vamanaṃ virecanaṃ uddhavirecanaṃ adhovirecanaṃ sīsavirecanaṃ kaṇṇatelaṃ nettatappaṇaṃ natthukammaṃ añjanaṃ paccañjanaṃ sālākiyaṃ sallakattikaṃ dārakatikicchā mūlabhesajjānaṃ anuppādānaṃ* (? *anuppāyanaṃ*) *osadhīnaṃ paṭimokkho....* Buddhaghosa explains *osadhīnaṃ paṭimokkho* as the administering of medicines such as alkalines and then expelling them after the time appropriate to them has passed (*khārādīni datvā tadanurūpe khaṇe gate tesaṃ apanayanam*) Sv 1, 98). T. W. Rhys Davids's rendering "Administering medicines in rotation" and explanation "It is when, for instance, a purgative is first given and then a tonic to counteract the other, to set free its effect" miss the mark (see DB I, 26 and n. 1). Cf. the *Tevijja Sutta of the Dīgha Nikāya* (DN 1, 250–52), where these practices are repeated. In fact, the list is repeated in the first thirteen sections of the *Dīgha Nikāya*. Similarly at MN 1, 510–11 (*Māgandiyasutta*) in treating eye disease, physicians (*bhisakka*, Skt. *bhiṣaj*) and surgeons (*sallakatta*, Skt. *śalyakarttṛ*) prepare medicines (*bhesajja*) used in purges of the upper and lower parts of the body (*uddhavirecana, adhovirecana*), in collyria (*añjana*), in ointments (*paccañjana*), and in nasal therapy (*natthukamma*) (cf. Buddhaghosa: *uddhavirecanaṃ adhovirecanaṃ añjanaṃ paccañj*[?]*anādibhesajjaṃ kareyya*; at Ps 3, 219); and cf. AN 5, 218–19, at p. 30 above
25. See Chapters 5 and 6, 74–116, passim.
26. *Geography* 15.1.60. Augustus Meineke, ed., *Strabonis Geographica* (1877; reprint, Graz; Akademische Druck-u. Verlagsanstalt, 1969), 3: 993–94. Cf. also H. L. Jones, ed. and trans., *The Geography of Strabo*, vol. 7 (London: Heinemann, 1930), 102–5 (Loeb edition); H. C. Hamilton, trans., *The*

Geography of Strabo (London: Henry G. Bohn, 1857), 3: 110–11; and J. W. McCrindle, trans., *Ancient India as Described in Classical Literature* (1901; reprint, New Delhi: Oriental Books Reprint, 1979), 67–68.

27. CaSū 11.54.
28. Cf. CaSū 11.55; 16.34–36; 27.3, 349–50.
29. See especially CaSū 27–28 and SaSū 46; cf. also Chapter 5 on Materia Medica, and Chapter 6 on Treatments, 74–116, passim.
30. CaŚā 4.13; 5.3.
31. This approach is not unlike that of ancient Greek medicine (see Chattopadhyaya, *Science and Society in Ancient India*, 51–52, 60–80).
32. See CaSū 27.3; Vi 3.36; cf. CaSū 10.6.
33. Chattopadhyaya, *Science and Society in Ancient India*, 84.
34. Zimmermann, *The Jungle and the Aroma of Meats*, 133 (French, 149).
35. SN 4, 230–31; cf. BKS 4, 154–56. The formula is repreated at AN 2, 87 and AN 3, 131, where one who is free from illness (*appābādha*) is said to be one in whom these eight do not exist; cf. Nidd I, 370. See also Miln 134–38, where Nāgasena discusses the eight causes of suffering in the context of the Buddha's lack of sin.
36. See Chapter 4, 60, 65–66.
37. This is found in the *Tikicchakasutta* and *Vamanasutta* (AN 5, 218–19). Cf. AN 4, 320, where bile, phlegm, and cutting wind are among the causes of death. See also DN 2, 14, 293–94; MN 1, 57–59, and 3.90; AN 3, 23–24; and Khp 2, where bile and/or phlegm are mentioned.
38. On *ṛtu*, see in particular CaSū 6; Ni 1.12, 28; Vi 1.111, 3.4; Śā 2.45; Ci 8.179; SuSū 6. On *viṣama*, see CaNi 1.19, 22, 28; Vi 6.12; Śā 1.109, 8.30; Ci 3.295–96a, 15.50, 25.20; SuSū 19.20, 31.30; Utt 39.63–74. On *karman*, see CaSū 1.49, 52, 25.18–19; Ni 7.19–20; Śā 1.116–17, 2.21, 44; Ci 9.16; SuUtt 40.163b–66a; cf. SuSū 31.30.
39. CaSū 20.3–4; SuSū 1.24–25.
40. Chattopadhyaya, *Science and Society in Ancient India*, 13, 179–88, 400–404.
41. CaSū 25.18–19. The only surviving treatise on *karmavipāka*, the diseases caused by past actions and their remedies by means of *prāyaścitta* (expiation), is the *Madanamahārṇava of Śrī Viśveśvara Bhaṭṭa*, ed. Embar Krishnamacharya and M. R. Nambiyar (Baroda: Oriental Institute, 1953).
42. Mitchell G. Weiss, "*Caraka Saṃhitā* on the Doctrine of Karma," in Wendy D. O'Flaherty, ed., *Karma and Rebirth in Classical Indian Traditions* (Berkeley: University of California Press, 1980), 90–115. It should be pointed out that Suśruta does discuss *karman* as an etiological category, but like Caraka only on the level of theory (SuUtt 40.163b–66a).
43. See K. R. Norman, *Pāli Literature* (Wiesbaden: Otto Harrassowitz, 1983), 110–11.
44. Miln 134–38.
45. See P. C. Bagchi, "A Fragment of the Kāśyapa Saṃhitā in Chinese," *Indian Culture* 9 (1942–43): 53–65; and Chapter 4, 67.

46. SuŚā 2, especially vv. 24–43.
47. *Geography* 15.1.70. Meineke, *Strabonis Geographica*, 3: 1001; Cf. also Jones, *Geography of Strabo*, 7: 122–25; Hamilton, *Geography of Strabo*, 3: 117–18; and McCrindle, *Ancient India as Described in Classical Literature*, 76.
48. See Chapter 6 on Treatments, and Zysk, "Mantra in *Āyurveda*," 123–43.
49. CaSū 11.54.
50. CaSū 30.21. See, p. 25 above.
51. See, in particular, G. J. Meulenbeld, *The Mādhavanidāna and Its Chief Commentary*, Chapters 1–10 (Leiden: Brill, 1974), 403–6; and Filliozat, *La Doctrine classique de la médicine indienne*, 13–14 (English, 16–17).
52. Chattopadhyaya, *Science and Society in Ancient India*, 29–30, 172, 260–61, 323.
53. See especially CaSū 25 and Vi 8.3, 27.
54. Found at DN 22.4–5 (2, 293–94), the Pāli anatomical terms (with their Sanskrit equivalents) are in order as follows: *kesa* (*keśa*), *loma* (*roman*), *nakha* (*nakha*), *danta* (*danta*), *taca* (*tvac*), *maṃsa* (*māṃsa*), *nahāru* (var. *nhāru*) (*snāyu*), *aṭṭhi* (*asthi*), *aṭṭhimiñja* (*asthimajjan*), *vakka* (*vṛkka*), *hadaya* (*hṛdaya*), *yakana* (*yakṛt*), *kilomaka* (*kloman*), *pihaka* (*plihan*, *plihaṇaka*), *papphāsa* (*pupphusa*), *anta* (*antra*), *antaguṇa* (*antraguda*), *udariya* (from *udara*), *karīsa* (*karīṣa*), *pitta* (*pitta*), *semha* (cf. *śleṣman*), *pubba* (*pūya*), *lohita* (*lohita*), *seda* (*sveda*), *meda* (*medas*), *assu* (*aśru*), *vasā* (*vasā*), *kheḷa* (*kheṭa*), *siṅghāṇika* (*siṅghāṇaka*), *lasikā* (*lasīkā*), *mutta* (*mūtra*), *pathavīdhātu* (*pṛthivīdhātu*), *āpodhātu* (*āpodhātu*), *tejodhātu* (*tejodhātu*), and *vāyudhātu* (*vāyodhātu*). The list illustrates a consistent and standardized anatomical nomenclature in the two Indic languages. It is repeated at MN 1, 57–59 (*Satipaṭṭhānasutta*), AN 3, 23–24 (*Udāyīsutta*), and Khp 2 (*Dvattiṃsākāra*); cf. MN 3, 90 (*Kāyagatāsatisutta*) and Sn 193–206.
55. Khap 2 (*matthaluṅga* is Sanskrit *mastuluṅga*). In one reading, *matthaluṅga* is inserted between *karīsa* and *pitta*; in another, it is found after *mutta: matthake matthaluṅgam.*
56. DN 22.6 (2, 294), etc.
57. DN 22.7–10 (2, 295–97), etc. Cf. *Suttanipāta*, 958, where the *bhikkhu*, who finds repose in a charnel ground, is extolled, and *Vaikhānasasmārtasūtra*, 8.9, where the Paramahaṃsa ascetic (*bhikṣu*) is defined as one who dwells, among other places, on a cremation ground (*śmaśāna*).
58. See Zysk, "Evolution of Anatomical Knowledge in Ancient India," 687–705. Cf. Chattopadhyaya, *Science and Society in Ancient India*, 94–100.
59. SuŚā 5.47–51.
60. A passage from the *Cullavagga* states that a corpse cast into a great ocean (i.e., large body of water) would very often be given up by the ocean and be forced ashore. Following this, there is a comparison between the great ocean and major rivers (CV 9.1.3). This is repeated at Miln 4.3.39 (187) and 4.6.33 (250). This suggests that the disposal of the dead in lakes and rivers was practiced early in India. There is, however, no evidence that Buddhists engaged in such practices.
61. Samuel Beal, trans., *Si–Yu–Ki*: *Buddhist Records of the Western World* (1884;

reprint, Delhi: Motilal Banarsidass, 1981), 2: 86; cf. Thomas Watters, trans., *On Yuan Chwang's Travels in India* (*A.D. 629–647*) (1904–5; reprint, New Delhi: Munshiram Manoharlal, 1973), 174.

62. Edward C. Sachau, trans., *Alberuni's India* (1910; reprint, New Delhi: Oriental Books Reprint, 1983), 2: 169.

Chapter 3

1. See S. B. Deo, *History of Jaina Monachism: From Inscriptions and Literature* (Puṇe: Deccan College Postgraduate and Research Institute, 1956), 209–10, 326–28, 437; and Hariprada Chakraborti, *Asceticism in Ancient India* (Calcutta: Punthi Pustak, 1973), 378, 384–401, 425.
2. CaSū 9.19. The theory was first proposed by Hendrik Kern, who stated that the Four Noble Truths (*āryasatyāni*) and "nothing else but the four cardinal articles of Indian medicine, applied to spiritual healing of mankind," citing Vedavyāsa's commentary (seventh to ninth centuries C.E.) on *Yogasūtra* 2.15 as proof (*Manual of Indian Buddhism* [reprint, Varanasi: Indological Book House, 1986, 46–47). He further cited two passages from the *Lalitavistara* (22 and 23.6, Vaidya's edition pp. 254, 258), which mention that the Buddha, after propounding the Four Noble Truths, appeared as a king among physicians (*vaidyarājan*), liberated from all suffering (*duḥkha*) and disease (*vyādhi*). These few references, according to Kern, established the connection between the Four Noble Truths and Indian medical science. All his citations are quite late, with no reference to the early medical or Buddhist sources. The passages from the *Lalitavistara* make no mention of an association between the four truths and medicine. Rather, the "king among physicians" (*vaidyarājan*) brings to mind the bodhisattva Bhaiṣajyaguru, also known in the *Saddharmapuṇḍarīka* (second century C.E.) as Bhaiṣajyarājan, Royal Physician, whose cult became popular in Central Asia, Tibet, and China. Moreover, Vedavyāsa's comments are to a Yoga treatise, not to a Buddhist one, and are not based on any division of medical science explicit in the classical treatises of Caraka (Bhela) and Suśruta. Cf. also Albrecht Wezler, "On the Quadruple Division of the Yogaśāstra, the Caturvyūhatva of the Cikitsāśāstra and the 'Four Noble Truths' of the Buddha," *Indologica Taurinensia* 12 (1984): 290–337, in which he critically investigates the interrelationship of fourfold divisions in Yoga, medicine, and Buddhism. He shows that the Buddhist use of the fourfold medical division first occurs in the fourth-century C.E. Mahāyāna treatise *Yogacarābhumi*. It closely resembles the passage from Caraka and probably derived from the medical tradition. He asserts that although certain Hīnayānists evidently knew the fourfold division of medicine and compared to it with the Four Noble Truths, the medical analogy cannot be traced back to the Buddha. He concludes that the Buddha did not borrow from medicine

to formulate his Four Noble Truths, but medical analogies were used in later Buddhist sources for the sake of illustration. In spite of these serious difficulties, scholars blindly followed Kern's claim. They include Erich Frauwallner, *History of Indian Philosophy* (New York: Humanities Press, 1974), 1: 146; and, recently, Richard Gombrich, *Theravāda Buddhism* (London and New York: Routledge & Kegan Paul, 1988), 59. Wezler cites several other eminent scholars who accepted analogy without question ("On the Quadruple Division of the Yogaśāstra," 312–14).

3. For the history and development of the Buddhist *saṅgha*, the reader should consult Mohan Wijayaratna, *Le Moine bouddhiste selon les textes du Theravâda* (Paris: Les Editions du Cerf, 1983); and Sukumar Dutt, *Buddhist Monks and Monasteries of India* (London: George Allen and Unwin, 1962). The principal secondaiy sources on the evolution and development of Buddhist monasticism and monasteries include these works as well as S. Dutt, *Early Buddhist Monachism*, rev. ed. (New Delhi: Munshiram Manoharlal, 1964). See also Patrick Olivelle, *The Origin and the Early Development of Buddhist Monachism* (Colombo: M. D. Gunasena, 1974); Nalinaksha Dutt, *Early Monastic Buddhism*, 2d ed. (Calcutta: Firma K. L. Mukhopadhyay, 1971); Dipak Kumar Barua, *Vihāras in Ancient India: A Survey of Buddhist Monasteries*, Indian Publications Monograph Series, no. 10 (Calcutta: Indian Publications, 1969); Rabindra Bijay Barua, *The Theravāda Saṅgha*, The Asiatic Society of Bangladesh Publications, no. 32 (Dacca: Asiatic Society of Bangladesh, 1978); Gombrich, *Theravāda Buddhism*; James Heitzman, *The Origin and Spread of Buddhist Monastic Institutions in South Asia* 500 *B.C.–A.D.*, South Asia Regional Studies Seminar Student Papers, no. 1 (Philadelphia: Department of South Asia Regional Studies, University of Pennsylvania, 1980); and R. Spence Hardy, *Eastern Monachism: An Account of the Origins, Laws, Discipline, Sacred Writings, Mysterious Rites, Religious Ceremonies, and Present Circumstances, of the Order of Mendicants Founded by Gotama Budha* [*sic*] (London: Partridge and Oakey, 1850). The last work is antiquated and very much influenced by Christian monasticism. On the possible revised date of the Buddha, see Chapter 2, n. 21, p. 141.
4. Gombrich, *Theravāda Buddhism*, 18–19.
5. Heinz Bechert and Richard Gombrich, eds., *The World of Buddhism: Buddhist Monks and Nuns in Society and Culture* (London: Thames and Hudson, 1984), 81.
6. CaVi 8.13, 20. See p. 24 above.
7. MV 1.30.4, 77. Cf. *Vaikhānasasmārtasūtra* 8.9, where the Haṃsa ascetic (*bhikṣu*) is described as one who subsists on cow's urine and dung (*gomūtragomayāhāriṇa*), and the Paramahaṃsa ascetic is said to be one who dwells under a tree with a single root (*vṛkṣaikamūla*).
8. CV 10.17.8; 23.3. See also I. B. Horner, *Women Under Primitive Buddhism* (1930; reprint, Delhi: Motilal Banarsidass, 1975), 154–55.
9. MV 6.14.6.

10. CaSū 1.69; 14.4; Ci 10.41; SuSū 15.5, 11; 45.217–26.
11. *Suttavibhaṅga* 4.1.1; *Saṅgītisuttanta* 3.3 (DN 3, 268); SN 41.3, 4 (SN 4, 288, 291); and Nidd II, 523. Cf. Olivelle, *Origin and Early Development of Buddhist Monachism*, 60.
12. Sabbāsava Sutta 27 (MN 1, 10).
13. See Chapter 5, and MV 1.30.4.
14. MV 6.17.1–6, and especially 6.33.
15. CV 6.21.1–2 (= Parivāra 15.4–5); Pācittiya 13.3.1, 81.2.1.
16. See CV 6.21.3. Cf. R. B. Barua, *Theravāda Saṅgha*, 60.
17. See Chapters 5 and 6. Both the literature and the archaeological evidence show that necessary items were donated. See, in particular, MV 1.39.3, where monks openly solicited food and medicine for healing from the laity; *Ākaṅkheyya Sutta* 4–5 (MN 1, 33), which mentions that monks received clothing, food, lodging, medicine, and other necessities for the sick; and the Valabhī copperplate inscription of Dharasena (I) (588 C.E.), which mentions that clothing, food, lodging, and medicine in sickness were provided to the *saṅgha* for all monks who came from various quarters (D. B. Diskalkar, *Selections from Sanskrit Inscriptions* [*2nd cent. to 8th cent.* A.D.] [New Delhi: Classical Publishers, 1977], 110, 112).
18. MV 8.26.3.
19. These also occur at AN 3, 143–44, where *dhamma* (virtuous qualities) replaces *aṅga*; cf. GS 3, 110–11.
20. MV 8.26.6. Cf. BD 4, 432, n. 3.
21. MV 8.26.5.
22. MV 8.26.8.
23. MV 8.26.7. Cf. BD 4, 433, nn. 1, 2.
24. MV 8.27. See Paul Demiéville, "Byô," in *Hôbôgirin*, Troisième Fascicule (Paris: Adrien Maisonneuve, 1937), 236–38; English translation by Mark Tatz, *Buddhism and Healing* (Boston: University Press of America, 1985), 31–35, where parallels to the Chinese Buddhist canon are given.
25. See Chattopadhyaya, *Science and Society in Ancient India*, 325.
26. CaSū 9.6. Cf. also CaSū 1.126–33, where the physician is described as one who knows the principles governing the application of medicines, and the quack as one who does not know them; and CaSū 29.6–13, where numerous qualities of a good and bad physician are enumerated.
27. CaSū 9.8.
28. CaSū 9.9.
29. SuSū 34.19–20.
30. SuSū 34.21b–22a.
31. SuSū 34.24.
32. Horner, *Women Under Primitive Buddhism*, 333–34.
33. MV 6.23.1–3; AN 1.14.7 (1, 26). Cf. Horner, *Women Under Primitive Buddhism*, 334.
34. See Chapter 6 on Rectal Fistula, 114–16.

35. See, in particular, MV 1.39, which explains that people joined the *saṅgha* specifically to be treated by Jīvaka, resulting in the ordinance prohibiting the ordination of the sick. Cf. the analysis of his healings in Appendix I, 120–27. See also Chattopadhyaya, *Science and Society in Ancient India*, 327–28; and Demiéville, "Byô," 238 (English, 36).
36. *Sarvatta vijitamhi... dve cikīcha katā manussacikīchā ca pasucikīchā ca osuḍhāni ca yāni manussopagāni ca pasopagāni ca yatta yatta nāsti sarvatrā hārāpitāni ca ropāpitāni ca mūlāni ca phalāni ca yatta yatt[r]a nāsti sarvatta hārāpitāni ca ropāpitāni ca paṃthesū kūpā ca khānāpitā vracchā ca ropāpitā paribhogāya pasumanussānam.* The text of the Girnar rock edict is based on the edition of Jules Bloch, ed. and trans., *Les Inscriptions d'Aśoka* (Paris: Société d'Edition "Les Belles Lettres," 1950), 94, 95; cf. his translation and valuable notes, 93, 94, 95.
37. Cf. Julius Jolly, *Medicin* (Strassburg: Verlag von Karl J. Trübner, 1901), 16 (English, 19); and Demiéville, "Byô," 246 (English, 56).
38. See Heitzman, *Origin and Spread of Buddhist Monastic Institutions in South Asia*, which clearly shows from archaeological data that monasteries were situated close to trade routes.
39. See D. K. Barua, *Vihāras in Ancient India*, 62.
40. SN 4, 210–13 (cf. BKS 4, 142–45, and AN 3, 142–42 [cf. 3, 109–13]); cf. Demiéville, "Byô," 245 (English, 54).
41. D. C. Sircar has edited and commented on this inscription. He understands *vigatajvara* to refer to a Buddhist monk or to the Buddha as the best of monks. Thus *vigatajvarālaya* means either a residence of Buddhist monks, modifying the best monastery (*vihāramukkhya*), or a shrine (*ālaya*) of the Buddha, existing in the best monastery ("More Inscriptions from Nāgārjunikoṇḍa," *Epigraphia Indica* 35 [1963–64]: 17–18). Sircar's interpretation depends on the understanding of *vigatajvara* as referring to the Buddha or to a *bhikkhu*. Literally, it means "devoid of fever," but has the secondary sense of "devoid of mental distress," that is, mentally and spiritually sane. One Buddhist reference to *vigatajvara* lends support it as an epithet of the Buddha: in the Jātaka of the deer Śiriprabha found in the *Mahāvastu*, the Buddha speaks to a group of monks while being *vigatajvara*, *vigatabhaya* (without fear), and *aśoka* (without grief) (see Émile Senart, ed., *Le Mahāvastu* [1890; reprint, Tokyo: Meicho-Fukyu-Kai, 1977], 2: 237, l. 14–15). In this context, its probable meaning is "without mental distress"; Jones's rendering "without old age" is out of the question (J. J. Jones, trans., *The Mahāvastu* [London: Luzac, 1952], 2: 224). This is the only reference to the compound I could locate in Buddhist literature (cf. John Brough, ed., *The Gāndhārī Dharmapada* [London: Oxford University Press, 1962], 185–86). However, it also occurs as an epithet of the Lord Ātreya Punarvasu in the chapter on the etiology of fevers in the *Caraka Saṃhitā*: *vigatajvaraḥ bhagavān... punarvasuḥ* (CaNi 1.44). Here the rare compound, found in a medical treatise, applies to the semilegendary medical teacher Ātreya, who, according to the Buddhist tradition, was a renowned

Taxilan healer who instructed the famous physician Jīvaka Komārabhacca in the medical arts, especially the technique of opening the skull (see Chapter 4, 54–56). These three references to *vigatajvara* help to strengthen the connection between Buddhist monasticism and the medical tradition. Based on the reference in the *Caraka Saṃhitā*, the passage in the inscription could be interpreted as in the splendid chief monastic house, in the abode of the Lord Ātreya or in the shrine of Lord Ātreya. "The abode of Lord Ātreya" suggests a place where medicine was practiced according to the tradition of Ātreya, while "the shrine of Lord Ātreya" points to a temple devoted to the sage healer (see CaSū 1.6–14, 30–33), at which those monks who engaged in healing according to his tradition honored him and perhaps even treated patients. On the notion of shrines at Nāgārjunikoṇḍa, see Gregory Schopen, "On the Buddha and His Bones: The Conception of a Relic in the Inscriptions of Nāgārjunikoṇḍa," *Journal of the American Oriental Society* 108 (1988): 527–37.

42. See James Legge, trans., *The Travels of Fa-hien* (1886; reprint, New Delhi: Master Publishers, 1981), 79. Cf. Beal, *Si-Yu-Ki*, 1: lvii.
43. See A. S. Altekar and Vijayakanta Misra, *Report on Kumrahār Excavations 1951–1955* (Patna: K. P. Jayaswal Research Institute, 1959), 11, 41, 52–53, 103, 107, and pls. XXXII, no. 5; XXXIV B, no. 2; XXXV, nos. 4, 5, as well as the frontispiece.
44. See D. R. Regmi, *Inscriptions of Ancient Nepāl*, 3 vols. (New Delhi: Abhinav Publications, 1983), 1: 66; 2: 40.
45. See D. K. Barua, *Vihāras in Ancient India*, 120. At CaSū 15.6, it states that an auspicious house (*praśasta gṛha*) (i.e., type of hospice) should be provided, among other things, with a mortar (*udūkhala*; cf. Pāli *udukkhala*) and a pestle (*musala*). Cf. CaCi 1.16–24, where there is a description of a suitable place to undertake indoor treatment (*kuṭīpāveśika*).
46. See D. C. Sircar, *Epigraphical Discoveries in East Pakistan* (Calcutta: Sanskrit College, 1974), 35, 38.
47. See K. V. Subrahmanya Ayyar, "The Tirumukkūḍal Inscription of Vīrarājendra," *Epigraphia Indica* 21 (1931–32): 220–50. The establishment of medical institutions at the sites of southern Indian Hindu temples was evidently not uncommon (see S. Gurumurthy, "Medical Science and Dispensaries in Ancient South India as Gleaned from Epigraphy," *Indian Journal of History of Medicine* 5 [1970]: 76–79).
48. See A. S. Ramanath Ayyar, "Śrīraṅgam Inscription of Garuḍavāhana-Bhaṭṭa: Śaka 1415," *Epigraphia Indica* 24 (1937–38): 90–101. The *ārogyaśālā* and construction of the Dhanvantari shrine are mentioned in the late-fifteenth- or early-sixteenth-century history of the Vaisṇava Aḻvārs and Ācāryas entitled *Divyasūricarita, sarga* 17.86 (see T. A. Sampatkumarācārya and K. K. A. Veṅkatācāryi, eds., *Divyasūricaritam by Garuḍavāhana Paṇḍita* [Bombay: Ananthacarya Research Institute, 1978]. I thank Thomas Burke of Harvard University for bringing this reference to my attention.

49. See Chapter 4, 54–56; and Radha Kumud Mookerji, *Ancient Indian Education* (1947; reprint, Delhi: Motilal Banarsidass, 1960), 468–70.
50. Jā 4.171–75. See also Mookerji, *Ancient Indian Education*, 472.
51. See Mookerji, *Ancient Indian Education*, 332, 470, 477–91.
52. S. Dutt, *Buddhist Monks and Monasteries*, 211–13. See also Mookerji, *Ancient Indian Education*, 510.
53. S. Dutt, *Buddhist Monks and Monasteries*, 132–34.
54. Watters, *On Yuan Chwang's Travels in India*, 2: 164–69. See also Beal, *Si-Yu-Ki*, 2: 170–72. Cf. S. Dutt, *Buddhist Monks and Monasteries*, 328–48.
55. See Samuel Beal, trans., *The Life of Hiuen Tsang by Shaman Hwui Li* (London: Kegan Paul, Trench, Trübner, 1914), 112. Cf. S. Dutt, *Buddhist Monks and Monasteries*, 332–33.
56. Watters, *On Yuan Chwang's Travels*, 1: 155; and Beal, *Si-Yu-Ki*, 1: 78.
57. Watters, *On Yuan Chwang's Travels*, 1: 159–60, and Beal, *Si-Yu-Ki*, 1: 79.
58. J. Takakusu, trans., *A Record of the Buddhist Religion as Practised in India and the Malay Archipelago (A.D. 671–695) by I-Tsing* (1896; reprint, New Delhi: Munshiram Manoharlal, 1982), 127–28.
59. Ibid., 128–40. Cf. 167–85, on the science of grammar and philology (*śabdavidyā*) and on the science of logic (*hetuvidyā*). Caraka gives the eight limbs (*aṣṭāṅga*) of *āyurveda* as follows: *kāyacikitsā* (general medicine), *śālākya* (science of disease above and clavicle), *śalyāpahartṛka* (extraction of arrows; i.e., major surgery), *viṣagaravairodhikapraśamana* (cessation of poisons, artificial poisons, and poisons from disagreeable foods), *bhūtavidyā* (demonology), *kaumārabhṛtyaka* (pediatrics, including care of pregnant women and delivery of infants), *rasāyana* (science of elixirs), and *vājīkaraṇa* (science of aphrodisiacs) Sū 30.28). The first in the list, *kāyacikitsā*, indicates the emphasis of Caraka's approach to medicine. Suśruta's enumeration is the same but begins with *śalya* (major surgery), which is the emphasis of his medical discipline, and continues as follows: *śālākya*, *kāyacikitsā*, *bhūtavidyā*, *kaumārabhṛtya*, *agadatantra* (toxicology in general), *rasāyanatantra*, and *vājīkaraṇatantra* (Sū 1.7; cf. 1.8, where each branch or limb is defined). Cf. also Demiéville, who has the following harsh criticism of I-tsing's knowledge of Indian medicine: "Ce pèlerin se piquait de connaissances médicales: il avait étudié naguère la médecine, dit-il (Takakusu, *Record*, 128), mais y avait renoncé parce que ce n'était point là une profession 'correcte' [pour un Moine]; on ne sait jamais très bein si les méthodes thérapeutiques qu'il recommande sont empruntées à l'Inde ou à la Chine" ("Byô," 255 [cf. English, 76]; see also 260 [English, 89]).
60. Translation from Tibetan by Claus Vogel, "On Bu-ston's View of the Eight Parts of Indian Medicine," *Indo-Iranian Journal* 6 (1962): 290–91; cf. E. Obermiller, trans., *History of Buddhism (chos-hbyung) by Bu-ston*, 2 pts. (Heidelberg: Otto Harrassowitz, 1931, 1932), pt. 1, 48. Cf. S. Dutt. *Buddhist Monks and Monasteries*, 323.

Chapter 4

1. On the spread of Buddhism beyond India, see, in particular, David Snellgrove, *Indo-Tibetan Buddhism*, 2 vols. (Boston: Shambhala, 1987), 44, 324, 389, 409–45, 485; and Robinson and Johnson, *Buddhist Religion*, 107–213.
2. Snellgrove, *Indo-Tibetan Buddhism*, 118–48, 235, 360, 452–59.
3. Erich Frauwallner, *The Earliest Vinaya and the Beginnings of Buddhist Literature*, trans. L. Petech (Rome: Is. M.E.O., 1956). Cf. also Hajime Nakamura, *Indian Buddhism: A Survey with Bibliographical Notes* (Osaka: KUFS Publications, 1980), 315–18; and Akira Yuyama, *Vinaya-Texte*, in Heinz Bechert, ed., *Systematische Übersicht über die Buddhistische Sanskrit-Literatur* (*A Systematic Survey of Sanskrit Buddhist Literature*), pt. 1 (Wiesbaden: Franz Steiner Verlag, 1979).
4. Frauwallner, *Earliest Vinaya*, 4. In particular, the Vinayas of the Sarvāstivāda, Dharmaguptaka, Mahīśāsaka, and Theravāda schools closely resemble one another and derive from a common source. The Vinayas of the Mūlasarvāstivāda and the Mahāsāṁghika schools are for the most part in agreement with the others, but the former tends to contain more numerous and elaborated legends, while the latter is utterly different in its internal structure, pointing perhaps to a later reworking (ibid., 23–24, 42).
5. Ibid., 67.
6. The chapter on medicines occurs as Chapter 6 in the Theravāda (Pāli), Sarvāstivāda, and Mūlasarvāstivāda schools; Chapter 7 in the Dharmaguptaka; Chapters 7 and 8 in the Mahīśāsaka school; and in fragmentary form in the Vinaya of the Mahāsāṁghika school (ibid., 3, 91, 178, 180, 183, 185, 195, 200).
7. Ibid., 91–99. See also Jan Jaworski, who has examined the section of remedies in the Vinayas of the Mahīśāsaka and the Theravāda schools. He discovered that there is a general correspondence between the two but fundamental divergences in the section's overall structure ("La Section des remèdes dans le Vinaya des Malūśāsaka [*sic*] et dans le Vinaya pāli," *Rocznik Orjentalistyczny* 5 (1927): 92–101). Cf. also Jan Jaworski, "La Section de la nourriture dans le Vinaya des Mahīśāsaka," *Rocznik Orjentalistyczny* 7 (1929–30): 53–124, where he analyzes and translates the section on foods in the Vinaya of the Mahīśāsaka and finds, as in the case of the section on remedies, close correspondence between the Pāli and Chinese versions. He also notices that the distinction between remedies and foods is vague. Concerning the *Bhaiṣajyavastu* of the Mūlasarvāstivāda preserved in Sanskrit, Nalinaksha Dutt states: "Unfortunately the most important portions dealing with the use of medicines and the ecclesiastical rules relating to the use of medicines are missing and what we have in these leaves are mostly stories of the *Avadāna* type. In the first fourteen pages there is some information about medicines while in pp. 221–240 there are a few Vinaya rules relating to the acceptance by monks of molasses, meat, fruits, and uncooked food" (*Gilgit Manuscripts* [Calcutta: Calcutta Oriental Press, 1947], 3.1:1)
8. The stories of Jīvaka in the Mahīśāsaka Vinaya, translated by Buddhajīva in

423 or 424 C.E., occur at T 1421 in the Dharmaguptaka Vinaya; translated by Buddhayasas and Chu Fo-nien in 405 or 408 C.E., are found at T 1428; and in the Sarvāstivāda Vinaya, translated by Punyatara and Kumārajīva from 399 to 413 C.E., occur at T 1435. See also Frauwallner, *Earliest Vinaya*, 3, 97–98, 178, 183, 185, 195, 200.

9. Following are the sources of the Jīvaka legend: for the Pāli, Hermann Oldenberg, ed., *The Vinaya Piṭakam*, Vol. 1: *The Mahāvagga* (1897; reprint, London: Luzac, 1964), 268–89; I. B. Horner, trans., *The Book of the Discipline* (*Mahāvagga*) (London: Luzac, 1962), 4: 379–97; cf. T. W. Rhys Davids and Hermann Oldenberg, trans., *Vinaya Texts*, pt. 2 (1882; reprint, Delhi: Motilal Banarsidass, 1974), 171–95; and R. Spence Hardy, trans., *A Manual of Buddhism in Its Modern Development* (1853; reprint, Varanasi: The Chowkhamba Sanskrit Series Office, 1967), 237–49. For the Sanskrit-Tibetan, Nalinaksha Dutt, ed., *Gilgit Manuscripts* (Calcutta: Calcutta Oriental Press, 1942), 3.2: 23–52; S. Bagchi, ed., *Mūlasarvāstivādavinayavastu* (Darbhanga, India: The Mithila Institute, 1967), 1: 182–97; D. T. Suzuki, ed., *The Tibetan Tripiṭaka: Peking Edition* (Tokyo and Kyoto: Tibetan Tripiṭaka Research Institute, 1957), 41: 260–67; F. Anton von Schiefner, trans., 'Der Prinz Dshīvaka als König der Ärzte," *Mélanges Asiatiques tirés du Bulletin de l'Académie Impériale des Sciences de St.-Pétersbourg* 8 (1879): 472–514; and W. R. S. Ralston, trans., *Tibetan Tales Derived from Indian Sources, Translated from the Tibetan of the Kah-gyur by F. Anton von Schiefner* (London: Kegan Paul, Trench, Trübner, 1906), 75–109. For the Chinese, Edouard Chavannes, trans., "Sutra pronouncé par le Buddhu au sujet de l'Avadāna concernant Fille-de-Manguier (Āmrāpali) et K'i-yu (Jīvaka)" (no. 499), in *Cinq cents contes et apologues du Tripiṭaka chinois*, Vol. 3 (Paris: Ernest Leroux, 1911); cf. Vol. 2 (1911), 55–56, and Vol. 4 (1934), 246–47. Chavannes translates "Nai niu k'i yu yin yua king" (T 553, pp. 896–902). Another version entitled "Nai niu k'i p'o king" (T 554, pp. 902–6) occurs immediately after the previous one in the section of various *sūtra*s in the Chinese canon. Cf. also "Wen chi si yu tchong seng king" (T 701, pp. 802c–803) and T 1509. See Étienne Lamotte, trans., *Le Traité de a grande vertu de sagesse de Nāgārjuna* (Mahāprajñāpāramitāśāstra) (Louvain: Bureaux du Muséon, 1949), 2: 990–91 n.; and James Losang Panglung, *Die Erzählstoffe des Mūlasarvāstivāda-Vinaya: Analysiert auf Grund der Tibetischen Übersetzung* (Tokyo: The Reiyukai Library, 1981), 65. None of the complete versions of the Jīvaka legend appears to be found in the Vinaya sections of the Chinese canon, suggesting that the story of this famous physician might have been singled out and removed from the monastic code because of its popularity. See also Demiéville, "Byô," 261–62 (English, 92–93).

10. See Paul U. Unschuld, *Medicine in China: A History of Ideas* (Berkeley: University of California Press, 1985), 92–96. The practice of using acupuncture needles might even go back to the fifth century B.C.E., when the physician Pien Ch'io allegedly recommended the use of acupuncture to treat demon-related illnesses (ibid., 45).

11. Cf. G. P. Malalasekara, *Dictionary of Pāli Proper Names* (1937; reprint, New

Delhi: Munshiram Manoharlal, 1983), 1: 957 and nn. See also Jolly, *Medicin*, 68 (English, 84).

12. Cf. Chattopadhyaya, *Science and Society in Ancient India*, 338–41.
13. Filliozat, *La Doctrine classique de la médecine indienne*, 8–9 (English, 10–11).
14. CaSū 1.9; 25.24.
15. See Zysk, *Religious Healing in the Veda*, 67–68.
16. See Demiéville, "Byô," 242–43 (English, 47–48); and Raoul Birnbaum, *The Healing Buddha* (Boulder, Colo.: Shambhala, 1979), 117, 121, 125–26, 129, 136.
17. This case, although found in the Tibetan version, is wanting in the translation of Schiefner (and Ralston).
18. See F. Anton von Schiefner, "Mahākātjājana und König Tshaṇḍa-Pradjota," *Mémoires de l'Académie Impériale des Sciences de St.-Pétersbourg*, 7th series, vol. 22 (1875): 7–11; cf. Panglung, *Die Erzählstoffe des Mūlasarvāstivāda–Vinaya*, 182–83.
19. The account of Jīvaka's curing the Buddha derives from Vinayas of the Sarvāstivāda school (T 1535: 23.194b9–c11) and the Mahīśāsaka school (T 1421: 2.134a17–b20). It is wanting in Chavannes's translation from the *sūtra* section.
20. Demiéville, "Byô," 228–29, 249–57 (English, 9–12, 65–82).
21. Johannes Nobel, "Ein alter medizinischer Sanskrit-text und seine Deutung," *Supplement to the Journal of the American Oriental Society*, No. 11 (July–September 1951); and R. E. Emmerick, trans., *The Sūtra of Golden Light* (London: Luzac, 1970).
22. A. F. Rudolf Hoernle, ed. and trans., *The Bower Manuscript*, facsimile leaves, Nāgarī transcript, romanized transliteration and English translation with notes (Calcutta: Office of the Superintendent of Government Printing, India, 1893–1912).
23. See P. L. Vaidya, ed., *Saddharmapuṇḍarīkasūtra* (Darbhanga, India: Mithila Institute, 1960); and H. Kern, trans., *Saddharma-Puṇḍarīka or the Lotus of the True Law* (1884; reprint, New York: Dover, 1963). See also Snellgrove, *Indo-Tibetan Buddhism*, 60; and Nakamura, *Indian Buddhism*, 181, 186, 189.
24. See Birnbaum, *Healing Buddha*, 17–77.
25. See Nalinaksha Dutt, ed., *Bhaiṣajyagurusūtra in Gilgit Manuscripts* (Calcutta: Calcutta Oriental Press, 1939), 1: 54; cf. Birnbaum, *Healing Buddha*, 62. Gregory Schopen asserts that "empirical" medicine was known among the circle of Buddhists in the Gilgit area, but the actual practice seemed to focus on "karmic" medicine, that is, the healing of diseases caused by misdeeds in a former life; he also observes that medicine and healing are insignificant functions of Bhaiṣajyaguru. Karmic medicine could be the result of estoteric Tantric influences ("*Bhaiṣajyagurusūtra* and the Buddhism of Gilgit" [Ph.D. diss., Australian National University, 1979], 210–22).
26. Snellgrove, *Indo-Tibetan Buddhism*, 340.
27. See Ronald E. Emmerick, *A Guide to the Literature of Khotan* (Tokyo: The

Reiyukai Library, 1979), 46–49. Several of the fragments have been analyzed and rendered into French by Jean Filliozat, *Fragments de texts kotchéens de médecine et de magie* (Paris: Librairie d'Amérique et d'Orient, Adrien-Maisonneuve, 1948).

28. See Ronald E. Emmerick, ed. and trans., *The Siddhasāra of Ravigupta*, vol. 1: The Sanskrit Text; vol. 2: The Tibetan Version with Facing English Translation (Wiesbaden: Franz Steiner Verlag, 1980, 1982). Editions and translations of the Khotanese and Uighur versions are in preparation.
29. See Sten Konow, ed. and trans., "A Medical Text in Khotanese: Ch II 003 of the India Office Library," *Avhandlinger Utgitt av det Norska Videnskaps-Akademi I* (Oslo). II. Hist.-Filos. Klasse, 1940–41, no. 4, 49–104; and Emmerick, *Guide to the Literature of Khotan*, 48–49.
30. See Snellgrove, *Indo-Tibetan Buddhism*, 446–50.
31. A list of these texts, compiled from the *Bkaḥ-ḥgyu* (*Kanjur*) and the *Bstan-ḥgyur* (*Tanjur*), includes the following:

    1. *Sbyor-ba-brgya-pa* (Skt. *Yogaśataka*) of Klu-sgrub shabs (Nāgārjunapāda) translated by Jetakarṇa, Buddhaśrījñāna, N̄i-ma rgyal-mtshan bzaṅ-po.
    2. *Sman-ḥtsho-baḥi mdo* (Skt. *Jīvasūtra*) of Klu-sgrub sñiṅ-po (Nāgārjuna-garbha).
    3. *Slob-dpon klu-sgrub-kyis bśad-pa sman a-baḥi cho-ga* (Skt. *Ācāryanāgārjuna-bhāṣitāvabheṣajakalpa*) of Klu-sgrub (Nāgārjuna).
    4. *Sman-dpyad yan-lag brgyad-paḥi sñiṅ-poḥi ḥgrel-pa-las sman-gyi miṅ-gi rnam-graṅs shes-bya-ba* (Skt. *Vaidyakāṣṭāṅgahṛdayavṛttau bheṣajanāma-paryāyanāman*) of Zla-ba-la dgaḥ-ba (Candranandana).
    5. *Yan-lag brgyad-paḥi sñiṅ-poḥi rnam-par ḥgrel-pa tshig-gi don-gyi zla-zer shes-bya-ba* (Skt. *Padārthacandrikāprabhāsanamāṣṭāṅgahṛdayavivṛti*) of Zla-ba-la dgaḥ-ba (Candranandana), translated by Jārandhāra, Rin-chen-bzaṅ-po (Ratnabhadra).
    6. *Yan-lag brgyad-paḥi sñiṅ-po bsdus-pa shes-bya-pa* (Skt. *Aṣṭāṅgahṛ-dayasaṃhitānāman*) of Sman-pa chen-po Pha-khol (Mahāvaidya Vāgbhaṭa), translated by Jārandhāra, Rin-chen-bzaṅ-po (Ratnabhdra).
    7. *Yan-lag brgyad-paḥi sñiṅ-po shes-bya-baḥi sman-dpyad-kyi bśad-pa* (Skt. *Aṣṭāṅgahṛdayanāmavaiḍūryakabhāṣya*) of Pha-khol (Vāgbhaṭa), translated by Dharmaśrīvarman, Śākya blo-gros, and revised by Rig-pa gshon-nu, Dbyig-gi rin-chen.

    This information is compiled from two published catalogs of the Tibetan canon: D. T. Suzuki, ed., *The Tibetan Tripiṭaka: Peking Edition* (Tokyo and Kyoto: Tibetan Tripiṭaka Research Institute, 1961), 167: 818–20; and Hakuju Ui et al., eds., *A Complete Catalogue of the Tibetan Buddhist Canons* (Sendai, Japan: Tôhoku Imperial University, 1934), 659–60. See also Fernand Meyer, *Gso-ba rig-pa, le système médical tibétain* (Paris: Editions du Centre National de la Recherche Scientifique, 1981); Manfred Taube, *Beiträge zur Geschichte der medizinischen Literatur Tibets* (Sankt Augustin: VGH Wissenschaftsverlag,

1981), 10–25; and Claus Vogel, ed. and trans., *Vāgbhaṭa's Aṣṭāṅghṛdayasaṃhitā, The First Five Chapters of Its Tibetan Version* (Wiesbaden: Kommissionsverlag Franz Steiner, 1965), 18–36.

32. See Zimmermann, *The Jungle and the Aroma of Meats*, 213 (French, 232).
33. See Bhagawan Dash, ed. and trans., *Tibetan Medicine with Special Reference to Yoga Śataka* (Dharamasala: Library of Tibetan Works and Archives, 1976); Michael Schmidt, ed. and trans., "Das Yogaśata: Ein Zeugnis altindischer Medizir [sic] in Sanskrit und Tibetisch (Inaugural-dissertation zur Erlangung der doktorwürde vorgelegt der Philosophischen Fakultät der Rheinischen Friedrich-Wilhelms Univesität zu Bonn, 1978); Jean Filliozat, ed. and trans., *Yogaśataka, texte médical attributé à Nāgārjuna* (Pondichéry: Institut Français d'Indologie, 1979); and K. G. Zysk, "Review of Jean Filliozat, *Yogaśataka*," *Indo-Iranian Journal* 23 (1981): 309–13. Filliozat has identified and studied Khotanese fragments based on the *Yogaśataka* or translations of a lost Sanskrit commentary on it (*Fragments de texts koutchéens de médicine et de magie*, 31–48).
34. See Ronald E. Emmerick, "Sources of the *Rgyud-bźi*," *Zeitschrift der Deutschen Morgenländischen Gesellschaft*, suppl. III.2 (Wiesbaden: Franz Steiner, 1977), 1135–42. Cf. also his following articles: "Some Lexical Items from the *Rgyud-bźi*," in Louis Ligeti, ed., *Proceedings of the Csoma de Kőrös Memorial Symposium* (Budapest: Akadémiai Kiadó, 1978), 101–8; "A Chapter from the *Rgyud-bźi*," *Asia Major* 19 (1975): 141–62; "Tibetan *nor-ra-re*," *Bulletin of the School of Oriental and African Studies* 51 (1988): 537–39; and "Epilepsy According to the *Rgyud-bźi*," in G. J. Meulenbeld and Dominik Wujastyk, eds., *Studies on Indian Medical History* (Groningen: Egbert Forsten, 1987), 63–90. See also K. G. Zysk, review of G. J. Meulenbeld and Dominik Wujastyk, eds., *Studies in Indian Medical History*, *Indo-Iranian Journal* 32 (1989): 322–27; Meyer, *Gso-ba rig-pa*, 33–36, 80–101 passim; Taube, *Beiträge zur Geschichte der medizinischen Literatur Tibets*, 26–38; Rechung Rinpoche, *Tibetan Medicine* (Berkeley: University of California Press, 1976), especially the introduction by Marianne Winder, 1–28; Jean Filliozat, "Un Chapitre du *Rgyud-bźi* sur les bases de la santé et des maladies," in Jean Filliozat, *Laghu-Prabandhāḥ*: *Choix d'articles d'indologie* (Leiden: Brill, 1974), 233–42; Demiéville, "Byô," 243 (English, 50); and Vogel. On Bu-ston's Views of the Eight Parts of Indian Medicine," 290–95.
35. Demiéville, "Byô," 225–65 (English, 1–101). See also *Supplément au Troisème Fascicule du Hôbôgirin*, IV. Cf. Jean Filliozat, "La Médecine indienne et l'expansion bouddhique en Extrême-orient," *Journal Asiatique* 224 (1934): 301–7, which refers to both Tibet and eastern Asia.
36. See Chapter 2, 30.
37. Demiéville, "Byô," 228–29, 249–57 (English, 9–12, 65–82).
38. See Unschuld, *Medicine in China*, 144–48.
39. See chapter 6 on Eye Disease, 88–91.
40. P. C. Bagchi, "New Materials for the Study of the Kumāratantra of Rāvaṇa,"

Indian Culture 7 (1941): 269–86; cf. Jean Filliozat, "La Kumāratantra de Rāvaṇa," *Journal Asiatique* 226 (1935): 1–66.

41. See Meulenbeld, *Mādhavanidāna and Its Chief Commentary*, 395. The Sanskrit text has been edited and published: Hemarāja Śarmā and Śrī Satapāla Bhiṣagācārya, *The Kāśyapa Saṃhitā* (*or Vṛddhajīvakīya Tantra*) *by Vṛddha Jīvaka, Revised by Vātsya* (Varanasi: The Chaukhambha Sanskrit Sansthan, 1953, 1976).
42. See Bagchi, "Fragment of the Kāśyapa-Saṃhitā in Chinese", 53–64; cf. also Demviéville, "Byô," 257–62 (English, 82–92); and Nakamura, *Indian Buddhism*, 320.
43. See Chapter 2, 31.
44. A partial list of these with their canon references (*Taisho Issasikyo*) includes the following: *The Secret Method of Treating Maladies of Meditation* (T 620), *Sūtra to Enchant Tooth*[*ache*] (T 1326), *Sūtra to Enchant* [*Maladies of*] *the Eye* (T 1328), *Sūtra to Enchant* [*Maladies of*] *Infants* (T 1329), *Sūtra of Healing Hemorrhoids* (T 1325), *Sūtra of the Incantation That Destroys All Maladies* (T 1323), *Sūtra of the Incantation That Purifies All Eye Ailments* (T 1324), *Sūtra for Deliverance from Illness (T 2865), and Sūtra for the Protection of Life and Deliverance of Men from Illness, Suffering, and Danger* (T 2878). The last two are of Chinese or Korean origin. Another *sūtra* of indigenous (Taoist) origin is *Sūtra Spoken by the Buddha on Interrupted Cuisine* (cf. Demiéville, "Byô," 259–60 [English, 87–89]).
45. See Birnbaum, *Healing Buddha*, 77–124.

Chapter 5

1. Pācittiya 37.
2. MV 6.1.1–5. Part of this case is repeated at Nissaggiya 23.1.
3. VA 5, 1089.
4. This is a South Indian tree, the seeds of which contain about 40 percent fatty oil, called "bassia oil" (see Nadk 1, 181). Horner, referring to the plant in a footnote, prefers the unlikely meaning "oil with honey" (BD 2, 132).
5. Nissaggiya 23.2.
6. SuSū 45.97; CaSū 13.18; 27.231; cf. CaSū 25.40.
7. SuSū 45.92–93; CaSū 27.230.
8. SuSū 45.112–13; CaSū 27.286–94; cf. CaSū 13.13; 25.40.
9. SuSū 45.128–30; CaSū 27.286–87.
10. SuSū 45.132; CaSū 27.245; cf. CaSū 25.40
11. SuSū 45.159; CaSū 27.239; cf. P. V. Sharma, *Ḍalhaṇa and His Comments on Drugs* (New Delhi: Munshiram Manoharlal, 1982), 208; and Meulenbeld, *Mādhavanidāna and Its Chief Commentary*, 507.
12. CaSū 6.41–44; SuSū 6.11, 35–36; SuUtt 64.13b–21a; cf. CaSū 13.18, 20–21; and AHSū 3.49.

13. See MV 6.2.2, where the rules pertaining to cooking with fats at the proper time are expounded.
14. MV 6.2.1–2.
15. VA 3, 714.
16. CaSū 13.11; 25.295; Ci 28.128; cf. BhSi 8.27.
17. SuSū 45.131; cf. BhVi 1.15; Si 8.27.
18. See Zimmermann, *Jungle and the Aroma of Meats* (*La Jungle et le fumet des viandes*), Introduction and passim.
19. See, in particular, CaCi 28.128.
20. Variant: *vacatta*; comm (VA 5, 1090): *sesavaca* (remaining *vaca*), var. *setavaca*, Skt. *śvetavaca* (white *vaca*). According to Ḍalhaṇa (to SuSu 39.3), *śvetavacā* = *śveta*, which in turn is equivalent to *vacā*. It is likely that Pāli *setavaca* is the correct reading, perhaps referring to a variety of *vacā*, with a predominant white color.
21. MV 6.3.1. The same list occurs at Pācittiya 11.2.1 and is repeated by Buddhaghosa at VA 4, 833. At Pācittiya 35.3.1 (cf. also 37.2.1), solid food (*khādaniya*) and soft food (*bhojaniya*) are defined: solid food includes all foods except the five (kinds of) meals (*pañcabhojana*) and the foods (such as certain types of fruits) taken during a night watch (*yāmakālika*), (those such as the five medicinal foods taken) during seven days (*sattāhakālika*), and (those such as pure medicines and herbs taken) during life (*yāvajīvaka*). Soft foods are the five (kinds of) meals: boiled rice (*odana*), food made with flour (*kummāsa*), barley meal (*sattu*), fish (*maccha*), and meat (*maṃsa*) (cf. BD 2, 330, and nn.). Buddhaghosa enumerates the various solid foods under the following categories: roots (*mūla*), tuberous roots (or bulbs) (*kanda*), lotus roots (*muḷāla*), top sprouts (*matthaka*), leaves (*patta*), flowers (*puppha*), stones of fruits (*aṭṭhi*), food from what is ground down (i.e., flour, etc.) (*piṭṭha*), and gums (or resins) (*niyyāsa*) (VA 4, 832–38). Several of these are included as medicines (see also R. B. Barua, *Theravāda Saṅgha*, 141).
22. BD 2, 228, n. 2.
23. MV 6.3.2.
24. VA 5, 1090.
25. Cf. BD 4, 272, n. 1.
26. CaSū 1.74–79
27. CaCi 1.41–45; SuSū 38.66–77 and Ḍalhaṇa.
28. Variant (comm. at VA 4, 835): *phaggava* = *paggava*. Buddhaghosa (at VA 5, 1090) glosses *pakkava* as *latājāti*, and Ḍalhaṇa (at SuUtt 60.48) glosses *latā* as *priyaṅgu*, the perfumed cherry or a type of grass. *Paggava* occurs at Jā 2.105, where it is called *vallī*, a type of creeper.
29. MV 6.4.1. It also occurs at VA 5, 835.
30. CaSū 1.66; 4.6–7; 26.43; SuSū 42.3, 9, 10; and Cakrapāṇidatta to CaSu 4.1, 6.
31. CaSū 4.7; cf. Meulenbeld, *Mādhavanidāna and Its Chief Commentary*, 453.
32. CaSū 4.11; cf. CaSū 3.8–9, where *tulasī, paṭola*, and *nimba* are mentioned in the treatment of skin disease.

33. SuSū 38.64–65.
34. MV 6.5.1. This also occurs at VA 4, 835.
35. CaSū 1.73–74.
36. CaSū 4.5.
37. SuSū 46.249–80; cf. CaSū 28.105–6.
38. See, in particular, SuCi 1.113; 4.30; SuKa 2.5–7 and SuUtt 14.10; and Ca (Bh), passim.
39. See BD 3, 245, n. 4.
40. See Jā 6.529, where *vibhītaka*, *harīṭaka*, and *āmalaka* occur together, and BD 4, 273, n. 1.
41. The compilers of the PED guess that it could refer to the plant *gotravṛkṣa* (255). More plausible is *koṭhaphalā*, found in Caraka. At CaSi 11.12, it is used against morbid pallor (*pāṇḍu*), and at CaKa 4.3, it is cited as a synonym of *dhāmārgava*-cucumber. The entire chapter, CaKa 4.3, is devoted to the plant's medicinal properties, with special emphasis on its fruits.
42. MV 6.6.1.
43. CaSū 27.6, 125–65; SuSū 46.139–280; cf. CaSū 1.74, 80–85, where nineteen plants with useful fruits are enumerated, and CaSi 11 and BhSi 7, the chapters on the successful measure of fruit (in enemas) (*phalamātrasiddha*).
44. CaSū 4.5.
45. CaSū 2.9; Ci 1.41–47; SuSū 38.43–47.
46. VA 5, 1090.
47. A proper identification of *taka* (*takka*) is at present wanting.
48. VA 5, 1090.
49. See PED, 669. The word *sarjarasa* has come to refer to the resin of the *sarja* tree, of the *rālā̆* tree, and of the Sal tree (*śāla*) (see DhNi 3.111, KaiNi 1421, and Toḍ 33.25; 35.13). Originally, it seems only to have been the sap of the *sarja* tree, sometimes mentioned as a synonym of *rālā̆* and *śāla* (cf. P. V. Sharma, trans., *Caraka Saṃhitā* [Varanasi: Chaukhambha Orientalia, 1983], 2: 731).
50. MV 6.7.1. It is quoted by Buddhaghosa at VA 4, 837, under his discussion of *niyyāsa* (gums).
51. Nissaggiya 18.2.
52. VA 3, 690; cf. BD, 2, 102, n. 10.
53. Pācittiya 4; cf. BD, 3, 249.
54. SuSū 27.18; 46.12; Ci 13.4–6; cf. DhNi 3.81 and SoNi 1.423.
55. SuSū 38.64–65.
56. References to Buddhaghosa's commments occur at VA 5, 1090.
57. MV 6.8.1.
58. CaSū 1.88b–92a; cf. CaCi 13.127.
59. CaSū 27.300–4.
60. BhCi 5.41; *sauvarcala* and *saindhava* occur at BhCi 28.28.
61. SuSū 46.313–25. On salts in the āyurvedic tradition, see N. S. Mooss, "Salt in Āyurveda I," *Ancient Science of Life* 6 (1987): 217–37.

62. CaSū 27.300–4; Vi 8.141; Ci 13.134.
63. CaSū 27.303b; SuUtt 42.92; see also ASSū 12.31.
64. CaCi 13.134.
65. SuSū 37.14 and SuUtt 42.92.

Chapter 6

1. MV 6.9.1. It also occurs at MV 8.17. Commentary found at VA 4, 884.
2. Ibid.
3. Elsewhere in the Vinaya, dung (*chakana*) and yellow clay (*paṇḍumattikā*) were used to dye monks' robes (*cīvara*). Buddhaghosa here glosses the dung as that from cows, and explains that the clay is that having the color of copper (*tamba*) (VA 5, 1126). This caused the robes to have an unfavorable color (*dubbaṇṇa*), so the Buddha permitted the use of six dyes, which are derived from roots (*mūla*), from stems (*khandha*), from barks (*taca*), from leaves (*patta*), from flowers (*puppha*), and from fruits (*phala*). Buddhaghosa specifies the particular dyes that are to be used: all root dyes are proper, with the exception of turmeric (*haliddā*); all stem dyes are proper, with the exception of those giving a crimson color (*mañjeṭṭha*) and the *tuṅgahara* (?), which is the name of a throṇy tree (*kaṇṭakarukkha*) whose yellow-colored extract (*haritālavaṇṇa*) is a stem dye; all bark dyes, with the exception of the lodh tree (*lodda*, Skt. *lodhra*) and white mangrove (?) (*kaṇḍala*) trees; all leaf dyes, with the exception of fresh leaves (*allipatta*) and dark leaves (*nīlipatta*); all flower dyes, with the exception of the dhak tree (*kiṃsuka*, Skt. *kiṃśuka*) and the safflower (*kusumbha*); no fruit dyes whatsoever are proper (ibid.). Because the monks used cold water to dye (Buddhaghosa: 'undecocted dye," ibid.), a bad smell was produced. The Buddha therefore permitted a small dye pot (*cullarajanakumbhī*) in which to decoct (*pācitum*) the dye. Along with this, all other materials necessary for the preparation of dyes by decoction were allowed to the monks (MV 8.10.1–3; cf. also BD 4, 405–6).

   The use of dyes was clearly well known in early monastic circles. The application of dung, clay, and dye in the treatment of minor skin conditions was quite popular and suggests a sympathetic approach to healing, current from Vedic times, which involved the dyeing of the skin to remove white patches on the skin (see Zysk, *Religious Healing in the Veda*, 81–2, 217–21). See also the use of dung allowed to monks in the treatment of "Snakebite," 101–3.
4. VA 5, 1090.
5. MV 6.9.2–10.1; cf. BD 4, 274, n. 3.
6. MV 8.17; Pācittiya 90.1–2.
7. CaSū 20.14; Ci 12.91; SuNi 13.16; cf. AHUtt 31.11. Cf. also the notion of thick sores (*sthūlārus* or *sthūlāruṣika*), described by Suśruta as large (at the base), difficult to cure (following Ḍalhaṇa), stiff wounds located among the joints (SuNi 5.9a) and further characterized by Bhela as colorless, inflamed,

slimy, wrinkled (?), and suppurating (BhCi 6.32). Likewise, compare Suśruta's description of scabs (*kacchū*): small eruptions similar to *pāman* rash characterized by suppurating swellings, itching, and excessive heat, and located on the buttocks and hands, and feet (SuNi 5.14b–15a). These are treated by an application of medicinal plaster (SuCi 20.17–19).

8. CaCi 12.93; SuCi 20.7–8. The sweet drugs are enumerated in the *jīvanīyagaṇa* at AHSū 25.8; see CaŚā 8.4, and Ḍalhaṇa to SuCi 20.7–8.
9. CaSū 3.8–11.
10. Ibid.; cf. CaCi 7.62–68 and passim. See also BhIn 2.2 and SuSū 4.11, where decoctions are used to remove itching, and SuCi 10.10.
11. CaSū 3.29.
12. SuSū 22.8; SuUtt 3.5–7.
13. See Zysk, *Religious Healing in the Veda*, 75–77, 207–8.
14. Ibid., 81–2.
15. See, in particular, CaCi 7.57; SuCi 10.16–17, 20, and passim; cf. BhCi 14.5.
16. MV 6.10.2.
17. VA 5, 1090; cf. BD 4, 274, n. 6.
18. SuUtt 60.6, 14.
19. SuUtt 60.28–30, 37–56. See also Zysk, "Mantra in *Āyurveda*," 123–43.
20. Zimmermann, *The Jungle and the Aroma of Meats*, 179 (French, 198).
21. In the Indian medical tradition, this is known as an extract of the wood of the Indian barberry (*dāruharidrā*) (Dutt and King *Materia Medica of the Hindus*, 74, 108).
22. The medical tradition describes this *añjana* as white in color. "It is said to be produced in the bed of the Jamuna and other rivers.... It is used as a collyrium for eyes, but is considered inferior to the black [collyrium]" (ibid. 74). On the first three *añjanas*, see Meulenbeld, *Mādhavanidāna and Its Chief Commentary*, 435–40.
23. See Nadk 1, 1189, and 1260–62.
24. *Kālānusārī* (*kāḷānusāriya*) is either a synonym of *tagara* or the black creeper (*kṛṣṇasārivā*) (see Sharma, *Ḍalhaṇa and His Comments on Drugs*, 139, 143). The latter is noted for its use in the treatment of eyes. At AN 5, 21–22, the plant is known as the best of all the scented roots (*mūlagandha*). Woodward, based on Benfey's Dictionary, renders "black gum" (GS 5, 17, n. 1; cf. also SN 3.156 and 5.44 and MN 3.6. At Ap 1.323, it is a substance with which the Buddha was anointed).
25. MV 6.11–12. Commentary found at VA 5, 1090–91. Cf. CV 5.28.2 and Pācittiya 86.2. The latter explains that *añjanī* and *añjanisalākā* are made of bone (*aṭṭhi*), ivory (*hatthidanta*), or horn (*visāṇa*). See also Horner, BD 4, 275 nn., and 3, 89, n. 2, where it is pointed out that *añjanī* is glossed by the commentator on the *Majjhimanikāya as añjananālikā* (a collyrium tube) (see Ps 3.302–3; cf. Thī 772–73 and Tha 36). Elsewhere in the canon, *nālikā* occurs along with *añjana*, and both are forbidden to monks (DN 1.7, 66). Buddhaghosa understands *añjana* to be cosmetic (*alaṃkārāñjana*) and glosses

nālikā as "medicine tube" (*bhesajjanālikā*). It is clear from these and other occurrences of *añjana* (cf. MN 2.64–65 and Thī 411) that collyria and the apparatuses for applying and storing them were used both for cosmetic purposes and for the healing of diseases of the eyes. The Buddhist monks of this period were permitted the use of them only in the latter case. Any form of bodily adornment was strictly forbidden.

26. The expression *cakkhuroga* occurs elsewhere in the canon as one of the common physical problems (*pākaṭaparissaya*) (see, in particular, AN 5.110 and Nidd I 13, 17, 46, 252, 269, 361 ff., 370, 407, 435, 365).
27. See also BhSū 26.20 and CaCi 26.224–256a, under *akṣiroga* (eye disease).
28. SuUtt 18.51–52. At ASSū 32.6, a fourth that lubricates (*sneha*) is added.
29. At SuUtt 18.57b–58a.
30. SuUtt 18.60b–63; see also SuSū 7.14 and CaCi 26.248. At SuCi 35.12 and BhSi 6.3–4, the enema tube is discussed. It is described in similar language. In Bhela, tin (*trapū̆*) and bamboo (*vaṃśa*) are added to the list of substances from which the tubes or rods are made.
31. CaCi 26.224–225a; SuUtt 1–19; cf. BhSū 4.25.
32. SuUtt 45.32.
33. CaSū 5.15; BhCi 16.44; SuUtt 17.13; cf. also ASSū 12.72 and AHSū 2.5.
34. See Meulenbeld, *Mādhavanidāna and Its Chief Commentary*, 438–40.
35. Dutt and King, *Materia Medica of the Hindus*, 73–74, and n. 1.
36. See the chapters on eye diseases in Suśruta, Utt 1–19, especially chapters 13–16 (17), which detail surgical procedures in the treatment of eye diseases. Cf. Jolly, *Medicin*, 114–15 (English, 138–39); and P. Kutumbiah, *Ancient Indian Medicine* (1969; reprint, Bombay: Orient Longmans, 1974), 175–76.
37. Jā 4, 401–12. An illustration of the Jātaka story occurs at Bhārhut (see Sir Alexander Cunningham, *The Stūpa of Bhārhut* [London: Allen, 1879], pl. xlvii.2).
38. MV 6.13.1–2. Commentary found at VA 5, 1091. Cf. Pārājika 3.5.15, where a monk suffering from the same head affliction was given medicine through the nose (*natthuṃ adaṃsu*) and died. This was a grave offense.
39. CaSū 17.6–29 and CaSi 9.70–117; cf. also BhCi 21 and BhSi 2; SuUtt 25.
40. SuUtt 25.5.
41. CaSū 17.6–29; 19.3.
42. CaSi 9.71–86; SuUtt 25.
43. SuUtt 25.8–10a.
44. SuUtt 26.42; SuCi 40; CaSi 9.89–110.
45. CaSi 9.102–5; SuCi 40.25.
46. See n. 30, above.
47. BhSū 6.28f.; BhSi 6.3–4; CaSū 5.59–61; SuCi 35.12; 40.3–9; cf. SuCi 40.10–20 for variations on this technique.
48. MV 6.14.1–2; cf. Pārājika 3.5.16, where a certain monk who was ill (*gilāna*) was rubbed with oil (*telena abbhañjiṃsu*) and died. This was a grave offense. Elsewhere in the Pāli canon, the Buddha was afflicted with wind

(*vātehābādhik*[*t*]*a*). He was treated by the elder Upavana, who bathed him with hot water (*uṇhodaka*) and gave him hot water mixed with molasses (*phāṇita*) to drink. The Buddha's affliction of wind, then, was calmed (SN 1, 174–75; Tha 185; cf. Miln 134. See especially Norman's notes at *The Elders' Verses* I [London: Luzac, 1969], 161, 184). No treatment precisely corresponding to this one occurs in the early medical treatises. Baths with warm water and certain medicines mixed with warm water are occasionally mentioned. A specific remedy of molasses in warm water is, however, wanting, suggesting that it was a remedy that was part of the general śramaṇic repository of ancient Indian medical lore. Cf. sections on treatments in the medical texts: CaCi 28.75ff.; SuCi 4; BhCi 24.

49. CaSū 20.11–13; SuSū 33.3b–7 and SuNi 1.
50. CaCi 28.134–82; SuCi 4 passim; cf. BhCi 24, passim.
51. CaCi 28.181–82.
52. SuCi 4.29.
53. The Pāli word *bhaṅga* does not mean "hemp," as Horner suggests (BD 4, 278–79, and n.). In the early medical texts, the term is found in the context of sweating or fomentations (*sveda*) and refers to sprouts, twigs, or buds of certain trees that are wind destroying. At SuUtt 17.62, the sweating of the eye should be treated with "sprouts which are destructive of wind" (*bhaṅgair anilanāśanaiḥ*). Ḍalhaṇa glosses *bhaṅga* with *pallava* (sprout), (cf. SuUtt 11.15, where Ḍalhaṇa glosses it with *patrabhaṅga* [leaves and sprouts] and says that in another place it is *pallava*). *Bhaṅga* in the Pāli, therefore, refers to those sprouts and leaves that destroy or remove wind (*vāta*).
54. MV 6.14.3. Commentary found at VA 5, 1091. Cf. Pārājika 3.5.14, where a certain monk who was ill (*gilāna*) was sweated (*sedesuṃ*) and died. This was a grave offense. Archaeological evidence from certain early Buddhist sites in Śrī Laṅkā indicates that special baths were constructed in Buddhist monasteries for the purpose of immersion therapy. Evidently, medicinal baths became a popular Śrī Laṅkan medical treatment that owes its origins to the early Buddhist monastic medical tradition. Its popularity among Śrī Laṅkan Buddhists explains Buddhaghosa's familiarity with the various techniques of immersion therapy (see R. A. L. H. Gunawardana, "Immersion as Therapy: Archaeological and Literary Evidence on an Aspect of Medical Practice in Precolonial Sri Lanka," *Sri Lanka Journal of the Humanities* 4 [1978]: 35–49).
55. CaCi 28.25; cf. also CaCi 28.55, 91, 227; CaSū 14.3–4; cf. BhSū, passim, and BhSū 21.6; CaSū 14.24; cf. CaNi 1; CaCi 4.10.
56. CaSū 14.39–63; SuCi 32.2–32; BhSū 12.1f.; BhCi 15.65f.; BhSi 4.33.
57. Ḍalhaṇa: "with blossoms, beginning with *eraṇḍa* [castor], which destroys wind." See also n. 53, above.
58. SuCi 32.7b–9a.
59. CaSū 14.50–51.
60. BhSū 12.1–30.
61. CaSū 14.44; BhSū 12.19; SuCi 32.13.

62. CaSū 14.45; BhSū 12.22b–24a; SuCi 32.13.
63. BhSū 22.25–26.
64. MV 6.14.4. Commentary found at VA 5, 1091–92.
65. CaCi 38.33, 37; cf. CaCi 38.228; SuNi 1.27; CaCi 29, SuNi 1.40–46; SuCi 5.1–17; cf. SuSū 24.9.
66. CaCi 29.12; cf. SuCi 5.4.
67. SuSū 14.25–48; SuŚā 8.25–26; cf. BhCi 6.41 and CaCi 29.37–40.
68. SuCi 4.8; cf. CaCi 28.93; SuCi 4.7, 10, 11a; 5.7.
69. CaCi 29.35–36; cf. also CaCi 38.92.
70. MV 6.14.4.
71. VA 5, 1092.
72. SuNi 13.29; SuCi 20.19b–20.
73. CaSū 5.92. The compound *pādasphuṭana* occurs at ASSū 3.60.
74. SuCi 24.70b–71b.
75. PED 452, 496.
76. Cakrapāṇidatta at CaSū 1.91.
77. Cf. section on Affliction Wind, where *abbhañjana* is found in connection with wind disease, 92–93.
78. BD 4, 279; PED 387; PTC 3(1), 43.
79. Jā 5.376: *gandhodakena pāde dhovitvā śatapakātelena abbhañjayiṃsu*; cf. also DN 2.240; Jā 3.120; 4.396, 476; 5.379. The commentaries likewise understand two operations performed on the feet: a foot wash (*pādadhovana*), usually with scented water (*gandhodaka*), and a foot massage (*pādabbhañjana*) with sesame oil (*tela*). Buddhaghosa's emendation of *pajja* to *majja* is supported by certain variant readings to the passage cited in the Jātaka. His explanation and justification for reading *majja*, however, is baseless.
80. MV 5.5.
81. CaSū 5.100.
82. MV 6.14.4–5. Commentary found at VA 5, 1092.
83. MV 1.39; cf. Ap 270.
84. MV 1.76.1–2; Pācittiya 2.2; cf. AN 5, 110.
85. Pācittya 60; Parivāra 2.2.6; cf. BD 3, 359–60 and nn.
86. Quite frequently in the canon, *gaṇḍa* occurs with *roga* (disease) and *salla* (dart or arrow). They are found together in formulaic passages used to describe certain emotional, mental, sensual, or physical states that hinder spiritual progress—for example, "Passion [*ejā*] is disease; passion is swelling; passion is a dart" (DN 2.283; cf. SN 4.64; see also SN 3.167, 189; 4.202–3; Nidd I 53, 56, 277; Nidd II 62–63; AN 2.128; 3.311; 4.289–90, 386–87, 422–24; Spk 2.334; Sn 51; and Thī 491). The term is metaphorically defined by a formula used to describe the body (*kāya*): " 'Swelling,' monks, is an expression of this body, is composed of the four gross elements, is born of mother and father, is an accumulation of gruel and sour milk, is impermanent, and is subject to effacement, abrasion, dissolution and disintegration. 'The root of the swelling [*gaṇḍamūla*]' is an expression of craving [*taṇhā*]" (SN 4.83).

87. CaSū 28.13–14; cf. BhSū 11.9; CaSū 11.49; SuNi 9–11; SuCi 16–18.
88. CaCi 25; SuCi 1; cf. also BhCi 27.
89. CaCi 25.39–43, 55–60, 96, 101–7.
90. SuCi 1.8.
91. SuCi 1.40, 57–58a, 65–70a, 88–89.
92. CaCi 25.44; SuCi 1.27b–30a.
93. Pācittiya 40.3.3.
94. MV 6.14.6.
95. Pācittiya 85.4.
96. CV 5.6. Cf. BD 5, 148–49 nn. Elsewhere in the canon and in postcanonical literature, cases of snakebite occur and are treated in various ways. The Visavanta Jātaka (no. 69) tells the story of a Bodhisatta who, born into a family of physicians skilled in the cure of snakebite, practiced the form of healing for his livelihood. Once when a man was bitten, he was summoned. The physician had the snake that bit the man caught and then tried to coerce it, by threatening it with destruction in fire, into sucking out its own venom. The snake refused to cooperate but was not destroyed in the fire. The physician subsequently extracted the poison by means of herbs and the recitation of charms. The man eventually recovered. Likewise in the *Milindapañha*, a cure for snakebite involved the recitation of incantations (*mantapada*) in order to make the snake suck back its own venom (150, 152: cf. MQ I 210, 213, and nn.). The notion of having the snake extract its own venom from the victim is very old. A similar notion can be found in charms against snakebite in the *Atharvaveda* (5.13.4 and 10.4.26; cf. 7.88 [93].1).
97. CaCi 23; SuKa 4–5.
98. CaCi 23.35–37, 192–198; SuKa 5.3f.
99. See above n. 96.
100. See, in particular, MN 1.79; DN 1.167; Jā 1.390 and Miln 259; cf. MLS 1, 106, n. 3; BD 1, 232, n. 1; and MQ 2, 71, and n. 4. See also *Vaikhānasasmārtasūtra* 8.9, where the Haṃsa ascetic (*bhikṣu*) is said to have been one who subsisted on cow's urine and cow's dung (*gomūtragomayāhāriṇa*).
101. See Chapter 3, 40. Cf. I-tsing's discussion and interpretation of putrid medicines in the Buddhist tradition (Takakusu, *A Record of the Buddhist Religion*, 138–40 and nn.).
102. CaCi 23.250–53.
103. SuKa 5.17.
104. SuKa 6.3–7. See also SuSū11, on the preparation and uses of alkalis; cf. CaCi 23.93–104 and BhCi 5.40–47, where different formulations of the cure called *kṣārāgada* are given.
105. CaSū 1.92–105; SuSū 45.217.
106. CaCi 23.46–50.
107. See Zysk, "Mantra in *Āyurveda*," 128.
108. MV 6.14.6.
109. MV 6.14.7. Commentary found at VA 5, 1092.

110. CaCi 23.105–122; SuKa 1.28–51a, 77–85.
111. SuKa 1.3–17.
112. Jyotir Mitra wrongly states that *gharadinnaka* is identical with *madātyaya* (intoxication), found at CaCi 14 (*Critical Appraisal of Āyurvedic Material in Buddhist Literature*, 247).
113. CaCi 23.9–14; Suśruta on the first two at SuKa 2.3–4.
114. CaCi 23.14 and commentary; ASUtt 40.14–15; AHUtt 35.5–6.
115. CaCi 23.233–241; cf. SuNi 8.12 and SuKa 8.24.
116. MV 6.14.7. Commentary found at VA 5, 1092.
117. CaCi 15; BhCi 11; SuUtt 40.167–87.
118. CaCi 15.56–57; cf. also SuUtt 40.170–72.
119. See Meulenbeld, *Mādhavanidāna and Its Chief Commentary*, 619 (cf. also 219–29).
120. CaCi 15.58–59; SuUtt 40.176–77.
121. CaCi 15.194–95; SuUtt 40.178–82.
122. CaCi 15.141–45, 168–93; BhCi 11.12; 5.40–47.
123. SuUtt 40.178–82.
124. SuSu 11.7–8; see also SuUtt 42.40–46a.
125. MV 6.14.6. Commentary found at VA 5, 1092.
126. CaCi 16; SuUtt 44. This section is wanting in Bhela. See also K. G. Zysk, "Studies in Traditional Indian Medicine in the Pāli Canon: Jīvaka and *Āyurveda*," *Journal of the International Association of Buddhist Studies* 5 (1982): 75–76, 82.
127. CaCi 16.44ff.; SuUtt 44.14ff.
128. SuUtt 44.16a, 21a; cf. CaCi 16.69.
129. SuUtt 44.23b, 25b; cf. CaCi 16.65a.
130. CaCi 16.55–69, 75.
131. CaCi 15.58a, 68a.
132. MV 14.6.7.
133. Mitra identifies *chavidosa* with *śītapitta* (urticaria) (*Critical Appraisal of Āyurvedic Material in Buddhist Literature*, 247). This equation is without merit.
134. CaNi 5; CaCi 7; BhNi 5; BhCi 6, which is incomplete, SuNi 5; SuCi 9.
135. CaNi 5.3–4; CaCi 7.9–26; BhNi 5; BhCi 6.11, 18f.; SuNi 5.3–6.
136. CaCi 5.37–42; BhCi 6; SuCi 9.6
137. CaCi 7.84–96; cf. BhCi 6.62f.; SuCi 9.10–11a.
138. See, in particular, DhNi 3, which treats aromatic substances.
139. See SuSū 24.9 and commentary, and especially SuCi 9.3–4. Caraka at Sū 3.29a and Bhela at Sū 6.17 describe a medicine that, when rubbed on the skin, removes *tvagdoṣa*.
140. SuCi 9.3.
141. In the Veda, *kúṣṭha* is the name of the costus plant. In *āyurveda*, it is both a general expression for skin disease and the plant. The latter is mentioned as one of the ingredients in some pastes administered in the treatment of

cutaneous afflictions (*kuṣṭha*), illustrating the religious notions of associative or sympathetic magic that were part of early āyurvedic medicine.

142. The medical treatises speak of two types of soups (*yūṣa*): unprepared (*akṛta*), which are boiled without spices, and prepared (*kṛta*), which are boiled with spices (see Meulenbeld, *Mādhavanidāna and Its Chief Commentary*, 492–93).
143. MV 6.14.7. Commentary found at VA 5,1092.
144. CaKa 12.8.
145. SuCi 33.4, 19, 10, 26.
146. CaKa 1; BhKa 1; SuSū 43, 44; SuCi 33.
147. SuSū 44.14; cf. CaKa 1.26.
148. SuCi 33.11; cf. CaSi 12.6–7.
149. MV 6.16.3.
150. Pārājika 3.5.33.
151. VA 2, 478–79; cf. BD 1, 149, n. 3.
152. See BD 4, 287 n. 2. The Sanskrit equivalent of *tekaṭula* is *trikaṭu(ka)*, which designates the three pungent or sharp substances: black pepper, long pepper, and dried ginger. The Buddhist enumeration is entirely different.
153. MV 6.17.1–5.
154. Pārājika 2.7.45. Commentary found at VA 2, 391. Horner mistakenly reads *phāṇita* (molasses) for *sakkarā* (granulated sugar) (BD 1, 111, n. 1).
155. CaCi 13; BhCi 13; SuNi 7; SuCi 14.
156. CaCi 13.59–67; SuCi 14.5 is much the same as Ca.
157. CaCi 13.99a. The five medicinal roots are, according to SuSū 38.68, as follows: Bengal quince (*bilva*), headache tree (*agnimantha*), Indian calosanthes (*ṭiṇṭuka*), trumpet flower tree (*pāṭalā*), and white teak (*kāśmarī*).
158. CaKa 7.72–73, also ASKa 2.32.
159. CaKa 9.7–8a, also ASKa 2.46.
160. According to the commentator Ḍalhaṇa, this refers to the *vidārigandhā* group of drugs. The first group is enumerated at SuSū 38.4–5, in the chapter on drugs and their properties. Verse 5 states: "This group, beginning with *vidārigandhā*, removes bile and wind and destroys consumption [*śosa*], *gulma* [internal or abdominal tumours], crushing [pain] in the limbs [*aṅgamarda*], *ūrdhaśvāsa* [asthma?, lit. 'elevated breathing,' i.e., shallow breathing], and cough [*kāsa*]."
161. This group is enumerated at SuSū 38.68 (see above n. 157). Verse 69 states: "The great [group] of five roots is considered to be combined with a bitter taste, destructive of phlegm and wind, stimulating of the digestive fire, and easy to digest, and has sweetness as a secondary flavor."
162. The commentary gives three possible glosses for this plant: *viṭapakarañja*, *cirapoṭika*, and *kākajaṅghā*. See also Sharma, *Ḍalhaṇa amd His Comments on Drugs*, 174.
163. SuSū 44.35–40a; cf. CaKa 11.17 and BhKa 8.20.
164. See especially CaCi 13.117–18a; CaKa 1.12; 7.13, 72; 8.17; 9.7–8a, 17; 10.11;

11.17; 12.35, 39; SuSū 20.23; 21.21; 42.11, 213; SuCi 8.17b; 11.5; SuKa 5.18b (proscribed for one with snakebite poison); and SuUtt 33.3b; 42.92.

165. SuCi 8.38; SuUtt 42.34–35.
166. CaVi 1.18.
167. MV 6.20.1–4.
168. See especially CaCi 3.31, 85, 89–108, 129–31; SuUtt 39.27, 32, 35, 47, 59, 76, 84, 180, 187, 282–93.
169. CaCi 3.222–23, 253, 258; SuUtt 39.166, 172, 222, 229, 236, 246, 297, 309–10.
170. CaCi 3.260–66.
171. SuUtt 39.388–93.
172. SuUtt 47.54–66.
173. The treatment appears to be wanting in Bhela; cf. BhCi 1 and 2.
174. MV 6.22.1–4. Commentary found at VA 5, 1093–94.
175. SuNi 4; SuCi 8.
176. SuNi 4.3.
177. CaCi 12.96–97; SuCi 8.4.
178. SuCi 8.38b–52a; cf. BhCi 2.14 and 4.79–87.
179. SuCi 8.52a–53.
180. SuCi 1.8; CaCi 25.38–43.

Appendix 1

1. Zysk, "Studies in Traditional Indian Medicine in the Pāli Canon," 70–86.
2. MV 8.1.7–11, 13.
3. MV 8.1.16–18.
4. See Chapter 6 on Head Irritated by Heat (Head Disease), 91–92.
5. See Chapter 4, 56.
6. MV 8.1.15.
7. CaCi 12.96–97; 25.38–43; SuCi 1.8; 8.38b–53; see also Chapter 6 on Rectal Fistula, 115.
8. See Zysk, "Studies of Traditional Indian Medicine in the Pāli Canon," 74.
9. See Chapter 6 on Rectal Fistula, 115.
10. MV 8.1.21–22.
11. Literally, the translation is: "...separates that which is other than the large intestines..."
12. The eleventh-century C.E. commentator Gayadāsa explains: "When bound for a long time, it appears, and when pressed down, it disappears."
13. SuNi 12.6.
14. The procedure is detailed at SuCi 19.48.
15. SuCi 19.20–24.
16. SuCi 2.56–66a; cf. CaCi 13.184b–188.
17. Mitra, *Critical Appraisal of Āyurvedic Material in Buddhist Literature*, 309–12. See also CaCi 13.39–41.

18. CaCi 13.184–188.
19. MV 8.1.23–25.
20. CaCi 16.44ff.; SuUtt 44.14ff.; see also Chapter 6 on Morbid Pallor or Jaundice, 106–7.
21. CaCi 26.50; here *pathyā* = *harītakī*.
22. Ḍalhaṇa suggests that it could also be understood as follows: "...he should drink clarified butter, *traiphala* [*ghṛta*], or *tailvaka* [*ghṛta*]." The last two are specific drugs, the principal ingredient of which is *ghṛta*, or clarified butter. On these see SuUtt 17.29 (cf. 10.14) and SuCi 14.10.
23. SuUtt 44.14–15.
24. MV 8.1.30–33.
25. CV 5.14.1–2.
26. CaKa 12.8.
27. SuCi 33.4, 19–20, 26; see also Chapter 6 on Body Filled (with the "Peccant" Humors), 108–10.
28. CaKa 1.19; SuSū 43.9.
29. SuSū 44.84b–86a. Translation follows Ḍalhaṇa.
30. CaKa 10.15–17.
31. Buddhaghosa explains: "Now, is the body of the Lord coarse? It is not coarse!...Divine beings always place the divine strength into the food of the Lord; and now the oily liquid moistens everywhere the humors; it makes the vessels supple" (Sv 1, 1118; see also Zysk, "Studies of Traditional Indian Medicine in the Pāli Canon," 77–78, 83 n. 56).
32. CaKa 1.4.
33. N. Dutt, ed., *Gilgit Manuscripts*, vol. 3.2, 47; cf. Bagchi, *Mūlasarvāstivāda-vinayavastu*, 1: 195. Suzuki, *Tibetan Tripiṭaka*, 41: 266–67 (leaves 66b–70a).
34. In the Mahīśāsaka, the nature of the Buddha's illness is not specified. It merely says that he suffered from a minor illness.
35. According the Mahīśāsaka, Jīvaka thinks: "I cannot use ordinary medicines on the Tathāgata, so I must use the medicine appropriate for a Cakravartin king."
36. Mahīśāsaka specifies that he should use three lotus (*uppala*) flowers.
37. The Sarvāstivāda version is found at T 1435:23.194b9–c11, and the Mahīśāsaka account occurs at T 1421:22.134a17–b20. I thank Dr. Robert Scharf of MacMaster University for his help in examining the Chinese passages.
38. CaKa 12.8.
39. SuCi 24.44b–46a.

Bibliography

Primary Sources (Texts and Translations)

Sanskrit Sources

Āgāśe, Kāśīnātha-Śāstrī, et al., eds. *Aitareyabrāhmaṇam: With the Commentary of Sāyaṇācārya.* 2 vols. Ānandāśrama Sanskrit Series, no. 32. Puṇe: Ānandāśrama, 1931.

———. *Kṛṣṇayajurvedīya-Taittirīya-Saṃhitā: With the Commentary of Sāyaṇācārya.* 8 vols. Ānandāśrama Sanskrit Series, no. 42. 1959–66. Reprint. Puṇe: Ānandāśrama, 1978.

Āṭhavale, Ananta Dāmodara, ed. *Vṛddhavāgbhaṭa's Aṣṭāṅgasaṃgrahaḥ with the Commentary of Indu.* Puṇe: Śrīmad Ātreya Prakāśanam, 1980.

Aufrecht, Theodor, ed. *Die Hymnen des Rigveda.* 2 vols. 1887. Reprint. Wiesbaden: Otto Harrassowitz, 1968.

Bagchi, S., ed. *Mūlasarvāstivādavinayavastu.* 2 vols. Buddhist Sanskrit Texts, no. 16. Darbhanga, India: The Mithila Institute, 1967, 1970.

Bandhu, Vishva, et al., eds. *Atharvaveda (Śaunaka) with the Padapāṭha and Sāyaṇācārya's Commentary.* 5 vols. Vishvesvaranand Indological Series, nos. 13–17. Hoshiarpur: Vishvesvaranand Vedic Research Institute, 1960–62.

Barret, LeRoy Carr, ed. "The Kashmirian Atharva Veda." *Journal of the American Oriental Society* 26 (1905): 197–295, 30 (1909–10): 187–258, 32 (1912): 343–90, 35 (1915): 42–101, 37 (1917): 257–308, 40 (1920): 145–69, 41 (1921): 264–89, 42 (1922): 105–46, 43 (1923): 96–115, 44 (1924): 258–69, 46 (1926): 34–48, 47 (1927): 238–49, 48 (1928): 34–65, 50 (1930): 43–73, 58 (1938): 571–64.

———. *The Kashmirian Atharva Veda: Books 16 and 17.* American Oriental Series, no. 9. New Haven, Conn.: American Oriental Society, 1936.

———. *The Kashmirian Atharva Veda: Books 19 and 20*. American Oriental Series, no. 18. New Haven, Conn.: American Oriental Society, 1940.

Bhattacharyya, Durgamohan, ed. *Atharvavedīyā Paippalāda Saṃhitā. Kāṇḍa 1*. Calcutta Sanskrit College Research Series, no. 26. Calcutta: Sanskrit College, 1964.

———. *Atharvavedīyā Paippalāda Saṃhitā: Kāṇdas 2–4*. Calcutta Sanskrit College Research Series, no. 62. Calcutta: Sanskrit College, 1970.

Bhishagratna, Kaviraj Kunjalal, trans. *An English Translation of the Sushruta Samhita Based on Original Sanskrit Text*. 3 vols. The Chowkhamba Sanskrit Series, no. 30. 1907–16. Reprint. Varanasi: The Chowkhamba Sanskrit Series Office, 1963.

Bloomfield, Maurice. *Hymns of the Atharvaveda*. Sacred Books of the East, no. 42. 1897. Reprint. Delhi: Motilal Banarsidass, 1964.

———. *The Kauśika Sūtra of the Atharvaveda, with Extracts from the Commentaries of Dārila and Keśava*. 1889. Reprint. Delhi: Motilal Banarsidass, 1972.

Caland, Willem, trans. *Altindisches Zauberritual, Probe einer Uebersetsung der wichtigsten Theile des Kauśika Sūtra*. 1900 (1936). Reprint. Wiesbaden: Martin Sändig, 1967.

———, ed. and trans. *Vaikhānasasmārtasūtram: The Domestic Rules and Sacred Laws of the Vaikhānasa School Belonging to the Black Yajurveda*. Bibliotheca Indica, nos. 242, 251. Calcutta: Asiatic Society of Bengal, 1927, 1929.

The Caraka Saṃhitā. Edited and Published in Six volumes with Translations in Hindī, Gujarātī, and English. Jamnagar: Shree Gulabkunverba Ayurvedic Society, 1949.

Dash, Bhagwan, and Lalitesh Kashyap, eds. and trans. *Materia Medica of Āyurveda: Based on Āyurveda Saukhyaṃ of Toḍarānanda*. New Delhi: Concept Publishing, 1980.

Dutt, Nalinaksha, ed. *Bhaiṣajyagurusūtra in Gilgit Manuscripts*. Vol. 1. Kashmīr Series of Texts and Studies, no. 71E. Calcutta: Calcutta Oriental Press, 1939.

———. *Gilgit Manuscripts*. Vol. 3.1, 2. Calcutta: Calcutta Oriental Press, 1942, 1947.

Edgerton, Franklin, ed. "The Kashmirian Atharva Veda: Book 6." *Journal of the American Oriental Society* 34 (1915): 374–411.

Eggeling, Julius, trans. *The Śatapatha-Brāhmaṇa according to the Text of the Mādhyandina School*. 5 vols. Sacred Books of the East, nos. 12, 26, 41, 43, 44. 1882–89. Reprint. Delhi: Motilal Banarsidass, 1963.

Elizarenkova, T. Ja., trans. *Atxarvaveda*. Moscow: Nauka, 1976.

Emmerick, R. E., trans. *The Sūtra of Golden Light*. London: Luzac, 1970.

———, ed. and trans. *The Siddhasāra of Ravigupta*. Vol. 1, The Sanskrit Text; Vol. 2, The Tibetan Version with Facing English Translation. Verzeichnis der Orientalischen Handschriften in Deutschland, suppl. 23.1, 2. Wiesbaden: Franz Steiner Verlag, 1980, 1982.

Filliozat, Jean, ed. and trans. *Yogaśataka: Text médical attribué à Nāgārjuna.*

Publications de l'Institute Français d'Indologie, no. 62. Pondichéry: Institut Français d'Indologie, 1979.

Geldner, Karl F., trans. *Der Rig-Veda.* 3 pts. Harvard Oriental Series, nos. 33–35. Cambridge, Mass.: Harvard University Press, 1951.

Gonda, Jan, trans. *The Savayajñas* (*Kauśika Sūtra 60–68*). Amsterdam: N. V. Noord-Hollandsche Uitgevers Maatschappi, 1965.

Griffith, Ralph T. H., trans *The Hymns of the Ṛgveda.* 5th ed. 2 vols. The Chowkhamba Sanskrit Series, no. 35. Varanasi: The Chowkhamba Sanskrit Series Office, 1971.

Henry, Victor, trans. *Les Livres VII–XII de l'Atharva-Veda.* Paris: J. Maisonneuve, 1891–96.

Hilgenberg, Luise, and Willibald Kirfel, trans. *Vāgbhaṭa's Aṣṭāṅgahṛdayasaṃhitā: Ein altindisches Lehrbuch der Heilkunde.* Leiden: Brill, 1941.

Hoernle, A. F. Rudolf., trans. *The Suśruta Saṃhitā or the Hindu System of Medicine According to Suśruta.* Fasciculus 1. Bibliotheca Indica, N. S., 911 (no. 139). Calcutta: Asiatic Society, 1897.

———, ed. and trans. *The Bower Manuscript.* Facsimile Leaves, Nāgarī Transcript, Romanized Transliteration and English Translation with Notes. Calcutta: Superintendent of Government Printing, India, 1893–1912.

Jones, J. J., trans. *The Mahāvastu.* Vol. 2. Sacred Books of the Buddhists, no. 18. London: Luzac, 1952.

Keith, Arthur B., trans. *The Veda of the Black Yajus School Entitled Taittirīya Sanhitā.* 2 pts. Harvard Oriental Series, nos. 18, 19. 1914. Reprint. Delhi: Motilal Banarsidass, 1967.

———. *Rigveda Brāhmaṇas: The Aitareya and Kauṣītaki Brāhmaṇas of the Rigveda.* Harvard Oriental Series, no. 25. Reprint. Delhi: Motilal Banarsidass, 1971.

Kern, H., trans. *Saddharma-Puṇḍarīka or the Lotus of the True Law.* Sacred Books of the East, no. 21. 1884. Reprint. New York: Dover, 1963.

Krishnamacharya, Embar, and M. R. Nambiyar, eds. *Madanamahārṇava of Śrī Viśveśvara Bhaṭṭa.* Gaekwad's Oriental Series, no. 117. Baroda: Oriental Institute, 1953.

Kunte, Anna Moresvaram, et al., eds. *Aṣṭāṅgahṛdayam Composed by Vāgbhaṭa with the Commentaries (Sarvāṅgasundarī) of Aruṇadatta and (Āyurvedarasāyana) of Hemādri.* 6th ed. Jaikrishnadas Āyurveda Series, no. 52. 1939. Reprint. Varanasi: Chaukhambha Orientalia, 1982.

Lamotte, Étienne, trans. *Le Traité de la grande vertu de sagesse de Nāgārjuna* (Mahāprajñāpārmitāśāstra). Vol. 2. Bibliothèque de Muséon, no. 18. Louvain: Bureaux de Muséon, 1949.

Meulenbeld, G. Jan, trans. *The Mādhavanidāna and Its Chief Commentary, Chapters 1–10.* Leiden: Brill, 1974.

Müller, F. Max, ed. *The Humns of the Rig-Veda with Sāyaṇa's Commentary.* 2d ed. 4 vols. The Chowkhamba Sanskrit Series, no. 99. 1890–92. Reprint. Varanasi: The Chowkhamba Sanskrit Series Office, 1966.

Pingree, David, ed. and trans. *The Yavanajātaka of Sphujidhvaja.* 2 vols. Harvard Oriental Series, vol. 48. Cambridge, Mass.: Harvard University Press, 1978.

Roth, R., and W. D. Whitney, eds. *Atharva-veda Saṃhitā*. Berlin: Ferd. Dümmlers Verlagsbuchhandlung, 1924.

Sampatkumarācārya, T. A., and K. K. A. Veṅkatācārya, eds. *Divyasūricaritam by Garuḍavāhana Paṇḍita*. Ananthacharya Research Institute Series, no. 2. Bombay: Ananthacharya Research Institute [1978].

Śarmā, Hemarāja, and Śrī Satyapāla Bhiṣagācārya, eds. *The Kāśyapa Saṃhitā or (Vṛddhajīvakīya Tantra) by Vṛddha Jīvaka, revised by Vātsya*. Kashi Sanskrit Series, no. 154. Varanasi: The Chowkhamba Sanskrit Series Office, 1953, 1976.

Śāstrī Nene, Gopāla, ed. *The Manusmṛti, with the "Manvarthamuktāvalī" Commentary of Kullūka Bhaṭṭa and the "Maṇiprabhā" Hindī Commentary by Haragovinda Śāstrī*. Kashi Sanskrit Series, no. 114. Varanasi: The Chowkhamba Sanskrit Series Office, 1970.

Senart, Émile, ed. *Le Mahāvastu*. Vol. 2. Société Asiatique. Collection d'ouvrages orientaux. Second series. 1890. Reprint. Tokyo: Meicho-Fukyu-Kai, 1977.

Sharma, E. R. Shreekrishna, ed. *Kauṣītaki-Brāhmaṇa*. Vol. 1. Wiesbaden: Franz Steiner Verlag, 1968.

Sharma, P. V., ed. *Soḍhala-Nighaṇṭu (Nāmasaṅgraha and Guṇasaṅgraha) of Vaidyācārya Soḍhala*. Gaekwad's Oriental Series, no. 164. Baroda: Oriental Institute, 1978.

———. *Kaiyadevanighaṇṭuḥ*. Translated into Hindī by Guru Prasad Sharma. Jaikrishnadas Āyurveda Series, no. 30. Varanasi: Chaukhambha Orientalia, 1979.

———. *Dhanvantarinighaṇṭuḥ*. Translated into Hindī by Guru Prasad Sharma. Jaikrishnadas Āyurveda Series, no. 40. Varanasi: Chaukhambha Orientalia, 1982.

———, ed. and trans. *Caraka-Saṃhitā: Agniveśa's Treatise Refined and Annotated by Caraka and Redacted by Dṛḍhabala*. 3 vols. Jaikrishnadas Āyurveda Series, 36.1–3. Varanasi: Chaukhambha Orientalia, 1981, 1983, 1985.

Singhal, G. D., et al., trans. *Ancient Indian Surgery*. [*Suśruta Saṃhitā*]. 12 vols. Varanasi: Singhal Publications, 1972–.

Schmidt, Michael, ed. and trans. "Das Yogaśata: Ein Zeugnis altindischer Medizir [*sic*] in Sanskrit und Tibetisch. Ph.D. diss., Friedrich-Wilhelms Universität, 1978.

Schroeder, Leopold von, ed. *Maitrāyaṇī Saṃhitā: Die Saṃhitā der Maitrāyaṇīya-Śākhā*. 4 vols. 1881–86. Reprint. Wiesbaden: Franz Steiner Verlag, 1970–72.

———. *Kāṭhakam: Die Saṃhitā der Kaṭha-Śākhā*. 3 vols. 1900–10. Reprint. Wiesbaden: Franz Steiner Verlag, 1970–72.

Tola, Fernando, trans. *Himno del Atharvaveda*. Buenos Aires: Editorial Sudamericana, 1968.

Trikamjī, Jādavjī, ed. *The Carakasaṃhitā by Agniveśa: Revised by Caraka and Dṛḍhabala, with the Āyurveda-Dīpakā Commentary of Cakrapāṇidatta*. 1941. Reprint. New Delhi: Munshiram Manoharlal, 1981.

Trikamjī, Jādavjī, and Nārāyaṇa Rāma Ācārya "Kavyatīrtha," eds. *The Suśrutasaṃhitā of Suśruta with the Nibandhasaṃgraha Commentary of Śrī*

Ḍalhaṇācārya and the Nyāyancandrikā Pañjikā of Śrī Gayadāsācārya. Jaikrishnadas Āyurveda Series, no. 34. 3d. ed. 1938. Reprint. Varanasi: Chaukhambha Orientalia, 1980.

Vaidya, P. L., ed. *Saddharmapuṇḍarīkasūtra.* Buddhist Sanskrit Texts, no. 6. Darbhanga, India: Mithila Institute, 1960.

Venkatasubramania Sastri, V. S., and C. Raja Rajeswara Sarma, eds. *The Bhela Saṃhitā.* New Delhi: Central Council for Research in Indian Medicine and Homoeopathy, 1977.

Vira, Raghu, ed. *Kapiṣṭhala-Kaṭha-Saṃhitā: A Text of the Black Yajurveda.* Delhi: Meharchand Lachmandas, 1968.

———. *Atharvaveda of the Paippalāda.* 1936–43. Reprint. Delhi: Arsh Sahitya Prachar Trust, 1979.

Vogel, Claus, ed. and trans. *Vāgbhaṭa's Aṣṭāṅgahṛdayasaṃhitā: The First Five Chapters of Its Tibetan Version.* Abhandlungen für die Kunde des Morgenlandes, 37.2. Wiesbaden: Kommissionsverlag Franz Steiner, 1965.

Weber, Albrecht, ed. *The Śatapatha-Brāhmaṇa in the Mādhyandina Śākhā with Extracts from the Commentaries of Sāyaṇa, Harisvāmin and Dvivedagaṅga.* The Chowkhamba Sanskrit Series, no. 96. 1855 Reprint. Varanasi: The Chowkhamba Sanskrit Series Office, 1964.

———, trans. "Erstes Buch–Fünftes Buch der Atharva-Saṃhitā." *Indische Studien* 4 (1858): 393–430, 13 (1873): 129–216, 17 (1885): 177–314, 18 (1898): 1–288.

Whitney, W. D., trans., and Charles R. Lanman, ed., *Atharva-veda-saṃhitā.* 2 pts. Harvard Oriental Series, nos. 7, 8. 1905. Reprint. Delhi: Motilal Banarsidass, 1971.

Pāli Sources

Andersen, Dines, and Helmer Smith, eds. *Sutta-nipāta.* Pāli Text Society. 1913. Reprint. Oxford: Geofrey Cumberlege (Oxford University Press), 1948.

Chalmers, Lord Robert, ed. *The Majjhima-nikāya.* Vols. 2, 3. Pāli Text Society, 1898–99. Reprint. London: Oxford University Press, 1951.

———, ed. and trans. *Buddha's Teachings: Being Sutta-Nipāta or Discourse-Collection.* Harvard Oriental Series, no. 37. Cambridge, Mass.: Harvard University Press, 1932.

Cowell, E. B., ed. *The Jātaka or Stories of the Buddha's Former Births.* Translated from the Pāli by various hands. 6 vols. 1895. Reprint. London: Pāli Text Society, 1973.

Fausböll, V., ed. *The Jātaka Together with Its Commentary, Being Tales of the Anterior Births of Gotama Buddha.* 7 vols. Indexes, compiled by Dines Andersen. Vol. 7, Pāli Text Society. 1877–97. Reprint. London: Luzac, 1962–64.

Feer, M. Léon, ed, *The Saṃyutta-nikāya of the Sutta-piṭaka.* 5 vols. Pāli Text Society, 1884–98. Reprint. London: Lunzac, 1960.

Hardy, E. ed. *The Aṅguttara-nikāya.* Vols. 3–5. Pāli Text Society, 1897–1900. Reprint. London: Luzac, 1958.
Hare, E. M., trans. *The Book of Gradual Sayings,* Vols. 3, 4. Pāli Text Society, Translations, nos 24, 25. 1934–35. Reprint. London: Luzac, 1952, 1955.
Horner, I. B., trans. *The Book of the Discipline.* 6 pts. Sacred Books of the Buddhists, nos. 10, 11, 13, 14, 20, 25. London: Luzac, 1938–52.
———. *The Collection of Middle Length Sayings* (*Majjhima-nikāya*). 3 vols. Pāli Text Society, Translations, nos. 29, 30, 31; 1954, 1957, 1959. Reprint. London: Luzac, 1967, 1970.
———. *Milinda's Questions.* 2 vols. Sacred Books of the Buddhists, nos. 22, 23. London: Luzac, 1963, 1964.
———, ed. *Papañcasūdanī: Majjhimanikāyaṭṭhakathā of Buddhaghosācārya.* Vols. 3–5. Pāli Text Society, 1933–38. Reprint London: Luzac, 1976–77.
Hunt, Mabel, ed., revised by C. A. F. Rhys Davids. *The Aṅguttara-nikāya.* Vol. 6, Indexes. Pāli Text Society, 1910. Reprint. London: Luzac, 1960.
Lilley, Mary E., ed. *The Apadāna of the Khuddakanikāya.* 2 vols. Pāli Text Society. London: Oxford University Press, 1925, 1927.
Morris, Richard, ed. *The Aṅguttara-nikāya.* 2 vols. Pāli Text Society, 1885, 1888. Reprint. London: Luzac, 1961, 1955.
Norman, K. R., trans. *The Elder's Verses. 1, Theragāthā; 2, Therīgāthā.* Pāli Text Society, Translations, nos. 38, 40. London: Luzac, 1969, 1971.
Oldenberg, Hermann, ed. *The Vinaya Piṭakam.* 5 vols. Pāli Text Society, 1879–83. Reprint. London: Luzac, 1964.
Oldenberg, Hermann, and R. Pischel, eds. *The Thera- and Therī-gāthā.* Pāli Text Society, 1883. Reprint. London: Luzac, 1966.
Poussin, L. de la Vallée, and E. J. Thomas, eds. *Niddesa I: Mahāniddesa.* 2 vols. Pāli Text Soceity. London: Oxford University Press (Humphrey Milford), 1916–17.
Rhys Davids, C. A. F., trans. *The Book of Kindred Sayings* (*Saṃyutta-nikāya*). 2 vols. Pāli Text Society, Translations, nos. 7, 10: 1922. Reprint. London: Luzac, 1950.
———. *The Majjhima-nikāya.* Vol. 4, Index of Words. Pāli Text Society, 1925. Reprint. London: Luzac, 1960.
———. *The Saṃyutta-nikāya of the Sutta-piṭaka.* Pt. 6, Indexes. Pāli Text Society, 1904. Reprint. London: Luzac, 1960.
Rhys Davids, T. W., trans. *The Questions of King Milinda.* 2 pts. Sacred Books of the East, nos. 35, 36. 1890, 1894. Reprint. New York: Dover, 1963.
Rhys Davids, T. W., and J. Estlin Carpenter, eds. *The Dīgha-nikāya.* 3 vols. Pāli Text Society, 1890–1911. Reprint. London: Luzac, 1960–67.
———. *The Sumaṅgala-vilāsinī: Buddhaghosa's Commentary on the Dīgha-nikāya.* Vol. 1. Pāli Text Society, 1886. Reprint. London: Luzac, 1968.
Rhys Davids, T. W., and Hermann Oldenberg, trans. *Vinaya Texts.* 3 pts. Sacred Books of the East, nos. 13, 17, 20. 1882, 1885. Reprint. Delhi: Motilal Banarsidass, 1974.

Rhys Davids, T. W., and C. A. F. Rhys Davids, trans. *Dialogues of the Buddha.* 3 pts. Sacred Books of the Buddhists, nos. 2–4. 1899–1921. Reprint. London: Luzac, 1956–66.

Smith, Helmer, ed. *The Khuddaka-pāṭha, Together with Its Commentary Paramatthajotikā*, Pt. 1. Pāli Text Society, 1915. Reprint. London: Luzac, 1959.

Stede, W., ed. *Niddesa II: Cullaniddesa.* Pāli Text Society, London: Oxford University Press (Humphrey Milford), 1918.

———. *The Sumaṅgala-vilāsinī: Buddhaghosa's Commentary on the Dīgha-nikāya.* Vols. 2, 3. Pāli Text Society, 1931–32. Reprint. London: Luzac, 1971.

Takakusu, J., and Makoto Nagai, eds. *Samantapāsādikā: Buddhaghosa's Commentary on the Vinaya Piṭaka.* 8 vols. Vol. 8, Indexes by H. Kopp. Pāli Text Society. London: Luzac, 1924–77.

Trenckner, V., ed. *Milindapañha.* Pāli Text Society, 1880. Reprint. London; Luzac, 1962.

———. *The Majjhima-nikāya.* Vol. 1. Pāli Text Society, 1888. Reprint. London: Luzac, 1964.

Woods, J. H., and D. D. Kosambi, eds. *Papañcasūdanī: Majjhimanikāyaṭṭhakathā of Buddhaghosācārya.* 2 vols. Pāli Text Society, 1922–28. Reprint. London: Luzac, 1977–79.

Woodward, F. L., ed. *Sāratthappakāsinī: Buddhaghosa's Commentary on the Saṃyutta-nikāya.* 3 vols. Pāli Text Society. London: Oxford University Press, 1939, 1932, 1937.

———, trans. *The Book of Kindred Sayings* (*Saṃyutta-nikāya*). Vols. 3–5: Pāli Text Society, Translations, no. 13, 14, 16: 1924, 1927, 1930. Reprint. London: Luzac, 1954, 1956, 1965.

———. *The Book of Gradual Sayings* [*Aṅguttaranikāya*]. Vols. 1, 2, 5. Pāli Text Society, Translations, nos. 22, 26, 27: 1932, 1933, 1936. Reprint. London: Luzac, 1960, 1962, 1961.

Other Sources

Beal, Samuel, trans. *Si-Yu-Ki: Buddhist Records of the Western World.* 2 vols. in 1. 1884. Reprint. Delhi: Motilal Banarsidass, 1981.

———. *The Life of Hiuen Tsang by Shaman Hwui Li.* London: Kegan Paul, Trench, Trübner, 1914.

Bloch, Jules, ed. and trans. *Les Inscriptions d'Aśoka.* Paris: Société d'Édition "Les Belles Lettres," 1950.

Brough, John, ed. *The Gāndhārī Dharmapada.* London Oriental Series, vol. 7. London: Oxford University Press, 1962.

Chavannes, Édouard, trans. *Cinq cents contes et apologues du Tripiṭaka chinois.* Vols. 1–4. Paris: Ernest Leroux, 1910–34.

Ebbell, B., trans. *The Papyrus Ebers.* Copenhagen: Levin & Muhksgaard, 1937.

Emmerick, Ronald E., ed. and trans. *The Siddhasāra of Ravigupta.* Vol. 1, The Sanskrit Text; Vol. 2, The Tibetan Version with Facing English Translation.

Verzeichnis der Orientalischen Handschriften in Deutschland, suppl. 23.1, 2. Wiesbaden: Franz Steiner Verlag, 1980, 1982.

Filliozat, Jean. *Fragments de texts koutchéens de médecine et de magie.* Texts, parallèles sanskrits et tibétains, traduction et glossaire. Paris: Librairie d'Amérique et d'Orient, Adrien-Maisonneuve, 1948.

Hamilton, H. C., trans. *The Geography of Strabo.* Vol. 3. London: Henry G. Bohn, 1857.

Jones, H. L., ed. and trans. *The Geography of Strabo.* Vol. 7. Loeb edition. London: Heinemann, 1930.

Konow, Sten, ed. and trans. "A Medical Text in Khotanese. Ch. II 003 of the India Office Library." *Avhandlinger Utgitt av det Norske Videnskaps-Akademi I Oslo. II. Hist.-Filos Klasse,* 1940, no. 4. 49–104.

Labat, René, ed. and trans. *Traité akkadien et diagnostics et pronostics médicaux.* Vols. 1, 2. Paris: Academie Internationale d'Histoire des Sciences; Leiden: Brill, 1951.

Legge, James, trans. *The Travels of Fa-hien.* 1886. Reprint. New Delhi: Master Publisher, 1981.

McCrindle, J. W., trans. *Ancient India as Described in Classical Literature.* 1901. Reprint. New Delhi: Oriental Books Reprint, 1979.

Meineke, Augustus, ed. *Strabonis Geographica.* Vol. 3. 1877. Reprint. Graz: Akademishe Druck-u. Verlagsanstalt, 1969.

Obermiller, E., trans. *History of Buddhism (chos-hbyung) by Bu-ston.* 2 pts. Materialien zur Kunde des Buddhismus, nos. 18, 19. Heidelberg: Otto Harrassowitz, 1931–32.

Ralston, W. R. S., trans. *Tibetan Tales Derived from Indian Sources, Translated from the Tibetan of the Kah-gyur by F. Anton von Schiefner.* London: Kegan Paul, Trench, Trübner, 1906.

Sachau, Edward C., trans. *Alberuni's India.* 2 vols. in 1. 1910. Reprint. New Delhi: Oriental Books Reprint, 1983.

Schiefner, F. Anton von, trans. "Mahākātjājana und König Tshaṇḍa-Pradjota: Ein Cyklus buddhistischer Erzählungen." *Mémoires de l'Académie Impériale des Sciences de St.-Pétersbourg,* 7th series, vol. 22 (1875): 1–67.

———. "Der Prinz Dsīvaka als König der Ärtze." *Mélanges Asiatiques Tirés du Bulletin de l'Académie Impériale des Sciences de St.-Pétersbourg* 8 (1879): 472–514.

Suzuki, D. T., ed. *The Tibetan Tripiṭaka: Peking Edition.* Vols. 41, 167, Tokyo-Kyoto: Tibetan Tripiṭaka Research Institute, 1957, 1961.

Takakusu, J., trans. *A Record of the Buddhist Religion as Practised in India and the Malaya Archipelago (A.D. 671–695) by I-tsing.* 1896. Reprint. New Delhi: Munshiram Manoharlal, 1982.

Takakusu, J., and K. Watanabe, eds. *Taisho Tripiṭaka: The Tripiṭaka in Chinese.* Tokyo: Taisho Isaikyo Kankawai, 1924–34.

Watter, Thomas, trans. *On Yuan Chwang's Travels in India (A.D. 629–645).* 1904–5. Reprint. New Delhi: Munshiram Manoharlal Publishers, 1973.

Secondary Sources

Allchin, Bridget, and Raymond Allchin. *The Rise of Civilization in India and Pakistan.* Cambridge World Archaeology. Cambridge: Cambridge University Press, 1985.

Altekar, A. S., and Vijayakenta Misra. *Report on Kumrahār Excavations, 1951–1955.* Patna: K. P. Jayaswal Research Institute, 1959.

Avaramuthan, T. G. *Some Survivals of the Harappan Culture.* Bombay: Karnatak,1942.

Ayyar, A. S. Ramanath. "Śrīraṅgam Inscriptions of Garudavāhana-Bhaṭṭa: Śaka 1415." *Epigraphia Indica* 24 (1937–38): 90–101.

Ayyar, K. V. Subrahmanya. "The Tirumukkūḍal Inscriptions of Vīrarājendra." *Epigraphia Indica* 24 (1931–32): 220–50.

Bagchi, P. C. "New Materials for the Study of the Kumāratantra of Rāvaṇa." *Indian Culture* 7 (1941): 269–86.

———. "A Fragment of the Kāśyapa Saṃhitā in Chinese." *Indian Culture* 9 (1942–43): 53–64.

Barua, Dipak Kumar. *Vihāras in Ancient India: A Survey of Buddhist Monasteries.* Indian Publications Monograph Series, no. 10. Calcutta: Indian Publications, 1969.

Barua, Rabindra Bijay. *The Theravāda Saṅgha.* The Asiatic Society of Bangladesh Publications, no. 32. Dacca: Asiatic Society of Bangladesh, 1978.

Basham, A. L. *History and Doctrines of the Ājīvikas; a Vanished Indian Religion.* London: Luzac, 1951.

———. *The Wonder That Was India: A Survey of the Culture of the Indian Sub-continent Before the Coming of the Muslims.* New York: Grove, 1959.

———. "The Background to the Rise of Buddhism." In *Studies in History of Buddhism*, edited by A. K. Narain, 13–17. Delhi: B. R. Publishing, 1980.

Bechert, Heinz. "The Date of the Buddha Reconsidered." *Indologica Taurinensia* 10 (1982): 29–36.

———. "A Remark on the Problem of the Date of Mahāvīra." *Indologica Taurinensia* 11 (1983): 287–90.

Bechert, Heinz, and Richard Gombrich, eds. *The World of Buddhism: Buddhist Monks and Nuns in Society and Culture.* London: Thames and Hudson, 1984.

Birnbaum, Raoul. *The Healing Buddha.* Boulder, Colo: Shambhala, 1979.

Bose, D. M., chief ed. *A Concise History of Science in India.* New Delhi: Indian National Science Academy, 1971.

Chakraborti, Haripada. *Asceticism in Ancient India, in Brāhmaṇical, Buddhist, Jaina and Ājīvaka Societies (from the Earliest Times to the Period of Śaṅkarācārya).* Calcutta: Punthi Pustak, 1973.

Chattopadhyaya, Debiprasad. *Science and Society in Ancient India.* Calcutta: Research India Publications. 1977.

———. *History of Science and Technology in Ancient India: The Beginnings.* Calcutta: Firma KLM, 1986.

Contenau, George. *La Médecine en Assyrie et en Babylonie*. Paris: Libraire Maloine, 1938.

Cunningham, Sir Arthur. *The Stūpa of Bhārhut: A Buddhist Monument Ornamented with Numerous Sculptures Illustrative of Buddhist Legend and History in the Third Century B.C.* London: Allen, 1879.

Dales, George F. "The Decline of the Harappans." In *Old World Archaeology: Foundations of Civilizations; Readings from Scientific American*, edited by C. C. Lamberg-Karlovsky, 157–64. San Francisco: Freeman, 1972.

Demiéville, Paul. "Byô." In *Hôbôgirin*, edited by Paul Demiéville, 224–65. Troisième Fascicule et Supplément. Paris: Adrien Maisonneuve, 1937. (Translated into English by Mark Tatz: *Buddhism and Healing*. Boston: University Press of America, 1985)

Deo, Shantaram Bhalchandra. *History of Jaina Monachism: From Inscriptions and Literature*. Puṇe: Deccan College Postgraduate Research Institute, 1956.

Diskalkar, D. B. *Selections from Sanskrit Inscriptions (2nd Cent. to 8th Cent. A.D.)*. New Delhi: Classical Publishers, 1977.

Dutt, Nalinaksha. *Early Monastic Buddhism*. 2nd ed. Calcutta: Firma K. L. Mukhopadhyay, 1971.

Dutt, Sukumar. *Buddhist Monks and Monasteries of India*. London: George Allen and Unwin, 1962.

———. *Early Buddhist Monachism*. Rev. ed. New Delhi: Munshiram Manoharlal, 1964.

Dutt, U. C., and George King et al. *Materia Medica of the Hindus*. Rev. ed. Calcutta: Madan Gopal Dass, 1922.

Eliade, Mircea. *Le Sacré et la profane*. Paris: Gallimard, 1965.

———. *The Quest: History and Meaning in Religion*. Chicago: University of Chicago Press, 1969.

———*The Myth of the Eternal Return*. Translated from the French by Willard R. Trask. Princeton N.J.: Princeton University Press, 1971.

———. *Shamanism: Archaic Techniques of Ecstasy*. Translated from the French by Willard R. Trask. Princeton, N.J.: Princeton University Press, 1972.

———. *Traité d'histoire des religions*. Paris: Petit Bibliothèque Payot, 1975.

Emmerick, Ronald E. "A Chapter from the *Rgyud-bźi*." *Asia Major* 19 (1975): 141–62.

———. "Sources of the *Rgyud-bźi*." *Zeitschrift der Deutschen Morganländischen Gesellschaft*. Suppl. III.2 (1977): 1135–41.

———. "Some Lexical Items from the *Rgyud-bźi*." In *Proceedings of the Csoma de Kőrös Memorial Symposium. Held at Mátrafüred, Hungary, 24–30 September, 1976*, edited by Louis Legeti, 101–8. Budapest: Akadémiai Kiadó, 1978.

———. *A Guide to the Literature of Khotan*. Studia Philologica Buddhica. Occasional Papers Series, vol. 3. Tokyo: The Reiyukai Library, 1979.

———. "Epilepsy According to the *Rgyud-bźi*." In *Studies on Indian Medical History*, edited by G. Jan Meulenbeld and Dominik Wujastyk, 63–90. Groningen: Egbert Forsten, 1987.

———. "Tibetan *nor-ra-re.*" *Bulletin of the School of Oriental and African Studies* 51 (1988): 537–39.

Fairservis, Walter A., Jr. *The Roots of Ancient India.* 2nd ed. Chicago: University of Chicago Press, 1975.

Filliozat, Jean. "La Médecine indienne et l'expansion bouddhique en Extrême-Orient." *Journal Asiatique* 224 (1934): 301–7.

———. "Le Kumāratantra de Rāvaṇa." *Journal Asiatique* 226 (1935): 1–66.

———. *La Doctrine classique de la médecine indienne; Ses origines et ses parallèles grecs.* Paris: Imprimerie nationale, 1949; Paris: Ecole Française d'Extrême Orient, 1975. (Translated into English by Dev Raj Chanana: *The Classical Doctrine of Indian Medicine.* New Delhi: Munshiram Manoharlal, 1964)

———. "Un Chapitre du *Rgyud-bźi* sur les bases de la santé et des maladies." In *Laghu-prabandhāḥ: Choix d'articles d'indologie*, 233–42. Leiden: Brill, 1974.

Frauwallner, Erich. *The Earliest Vinaya and the Beginnings of Buddhist Literature.* Serie Orientale Roma, no. 8. Translated by L. Petech. Roma: Is. M. E. O., 1956.

———. *History of Indian Philosophy.* 2 vols. Translated by V. M. Bedekar. Delhi: Motilal Banarsidass, 1974.

Gombrich, Richard F. "How the Mahāyāna Began." *Journal of Pāli and Buddhist Studies* 1 (1988): 29–46.

———. *Theravāda Buddhism: A Social History from Ancient Benares to Modern Colombo.* London and New York: Routledge & Kegan Paul, 1988.

Grapow, Hermann, *Grundiss der Medizin der alten Ägypter.* Vols. 1–4. Berlin: Akademie Verlag, 1954–59.

Gunawardana, R. A. L. H. "Immersion as Therapy: Archaeological and Literary Evidence on an Aspect of Medical Practice in Precolonial Sri Lanka." *Sri Lanka Journal of the Humanities* 4 (1978): 35–59.

Gurumurthy, S. "Medical Science and Dispensaries in Ancient South India as Gleaned from Epigraphy." *Indian Journal of History of Science* 5 (1970): 76–79.

Hardy, R. Spence, *Early Monachism: An Account of the Origins, Laws, Discipline, Sacred Writings, Mysterious Rites, Religious Ceremonies, and Present Circumstances of the Order of Mendicants Founded by Gotama Budha* [*sic*]. London: Partridge and Oakey, 1850.

———. *A Manual of Buddhism in Its Modern Development.* 1853. Reprint. Varanasi: The Chowkhamba Sanskrit Series Office, 1967.

Heitzman, James. *The Origin and Spread of Buddhist Monastic Institutions in South Asia, 500 B.C.–*300 *A.D.* South Asia Regional Studies Seminar, Student Papers, no. 1. Philadelphia: Department of South Asia Regional Studies, 1980.

Horner, I. B. *Women Under Primitive Buddhism.* 1930. Reprint. Delhi Motilal Banarsidass, 1975.

Jaworski, Jan. "La Section des remèdes dans le Vinaya des Mahūśāsaka [*sic*] et le Vinaya pāli." *Rocznik Orjentalistyczny* 5 (1927): 92–101.

———. "La Section de la nourriture dans le Vinaya des Mahīśāsaka." *Rocznik*

Orjentalistyczny 7 (1929–30): 53–124.

Jolly, Julius. *Medicin.* Grundriss der Indo-Arischen Philologie und Altertumskunde, no. 3.10. Strassburg: Verlag von Karl J. Trübner, 1901. (Translated into English by C. G. Kashikar: *Indian Medicine.* New Delhi: Munshiram Manoharlal, 1977).

Kern, Hendrik, *Manual of Indian Buddhism.* Grundriss der Indo-Arischen Philologie und Altertumskunde, no. 3.8. 1896. Reprint. Delhi: Indological Book House, 1968.

Kirby, E. T. *Ur-Drama: The Origin of Theatre.* New York: New York University Press, 1975.

Kuhn, Thomas S. *The Structure of Scientific Revolutions.* Chicago: University of Chicago Press, 1962.

Law, B. C. *The Life and Works of Buddhaghosa.* Calcutta: Thacker and Spink, 1923.

Macdonell, Arthur A. *Vedic Mythology.* Grundriss der Indo-Arischen Philologie und Altertumskunde, no. 3.1. 1898. Reprint. Delhi: Motilal Banarsidass, 1974.

Mackay, Ernest J. H. *Further Excavations at Mohenjo-Dāro, Being an Official Account of Archaeological Excavations at Mohenjo-Dāro Carried Out by the Government of India between the Years 1927 and 1931.* 2 vols. Delhi: Manager of Publications, Government of India Press, 1938.

———. *Chanhu-Dāro Excavations, 1935–36.* American Oriental Series, no. 20. New Haven, Conn.: American Oriental Society, 1943.

Malalasekara, G. P. *Dictionary of Pāli Proper Names.* 2 vols. 1937. Reprint. New Delhi: Munshiram Manoharlal, 1983.

Marshall, Sir John, ed. *Mohenjo-Dāro and the Indus Civilizations, Being an Official Account of Archaeological Excavations at Mohenjo-Dāro Carried Out by the Government of Inda between the Years 1922 and 1927.* 3 vols. London: Arthur Probsthain, 1931.

Meyer, Fernand. *Gos-ba rig pa, le système médical tibétain.* Paris: Editions du Centre Nationale de la Recherche Scientifique, 1981.

Mitra, Jyotir. *A Critical Appraisal of Āyurvedic Material in Buddhist Literature.* Varanasi: The Jyotirlok Prakashan, 1985.

Mookerji, Radha Kumud. *Ancient Indian Education.* 1947. Reprint. Delhi: Motilal Banarsidass, 1960.

Mooss, N. S. "Salt in Āyurveda I." *Ancient Science of Life* 6 (1987): 217–37.

Nadkarni, A. K. *Dr. K. M. Nadkarni's Indian Materia Medica.* 3d rev. ed. 2 vols. 1908. Reprint. Bombay: Popular Prakashan, 1976.

Nakamura, Hajime. *Indian Buddhism: A Survey with Bibliographical Notes.* Intercultural Research Institute Monograph, no. 9. Osaka: KUFS Publications, 1980.

Nobel, Johannes. "Ein alter medizinischer Sanskrit-text und seine Deutung." *Supplement to the Journal of the American Oriental Society* 11 (July–September 1951): 1–35.

Norman, Kenneth R. *Pāli Literature, Including the Canonical Literature in Prakrit and Sanskrit of All Hīnayāna Schools of Buddhism.* A History of Indian Literature, no. 7.2. Wiesbaden: Otto Harrassowitz, 1983.

Obeyesekere, Gananath. "Myth, History and Numerology in the Buddhist Chronicles." Forthcoming.

Olivelle, Patrick. *The Origin and the Early Development of Buddhist Monachism.* Colombo: M. D. Gunasena, 1974.

Panglung, Jampa Losang. *Die Erzählstoffe des Mūlasarvāstivāda-Vinaya: Analysiert auf Grund der Tibetischen Übersetzung.* Studia Philologica Buddhica, Monograph Series, no. 3. Tokyo: The Reiyukai Library, 1981.

Pingree, David. "Astronomy and Astrology in India and Iran." *Isis* 54 (1963): 229–46.

———. "The Mesopotamian Origin of Early Indian Mathematical Astronomy." *Journal for the History of Astronomy* 4 (1973): 1–12.

———. "The Recovery of Early Greek Astronomy from India." *Journal for the History of Astronomy* 7 (1976): 109–23.

———. "History of Mathematical Astronomy in India." In *Dictionary of Scientific Biography.* Vol. 15, edited by Charles Coulston Gillispie, 533–633. New York: Scribner, 1978.

Pollock, Sheldon. "The Theory of Practice and the Practice of Theory in Indian Intellectual History." *Journal of the American Oriental Society* 105 (1985): 499–519.

Possehl, Gregory L., ed. *Ancient Cities of the Indus Civilization.* Durham, N.C.: Carolina Academic Press, 1979.

Poussin, L. de la Vallée. *Histoire du monde.* Vol. 3, *Indo-européens et Indo-iraniens; l'Inde jusque vers 300 av. J.-C.* Paris: Editions de Boccard, 1924.

Rao, S. R. *Lothal and the Indus Civilization.* New York: Asia Publishing House, 1973.

Regmi, D. R. *Inscriptions of Ancient Nepāl.* 3 vols. New Delhi: Abhiv, 1983.

Renfrew, Colin. *Archaeology and Language: The Puzzle of Indo-European Origins.* Harmondsworth: Penguin Books, 1989.

Rhys Davids, T. W., and William Stede. *The Pāli Text Society's Pāli–English Dictionary.* 1921–25. Reprint. London: Pāli Text Society, 1972.

Rinpoche, Rechung. *Tibetan Medicine.* Berkeley: University of California Press, 1976.

Robinson, R. H., and W. L. Johnson. *The Buddhist Religion: A Historical Introduction.* 3d ed. Belmont, Calif.: Wadsworth, 1982.

Schopen, Gregory. "*Bhaiṣajyagurusūtra* and the Buddhism of Gilgit." Ph.D. diss., Australian National University, 1979.

———. "On the Buddha and His Bones: The Concept of Relic in the Inscriptions of Nāgārjunikoṇḍa." *Journal of the American Oriental Society* 108 (1988): 527–37.

Sharma, P. V. *Ḍalhaṇa and His Comments on Drugs.* New Delhi: Munshiram Manoharlal, 1982.

Sigerist, Henry E. *A History of Medicine.* Vol. 1, *Primitive and Archaic Medicine.* New York: Oxford Univesity Press, 1955.

Sircar, D. C. "More Inscriptions from Nāgārjunikoṇḍa." *Epigraphia Indica* 35 (1963–64): 17–18.

———. *Epigraphic Discoveries in East Pakistan.* Calcutta: Sanskrit College, 1974.

Sivin, Nathan. "Science and Medicine in Imperial China—The State of the Field." *Journal of Asian Studies* 47 (1988): 41–90.

Snellgrove, David. *Indo-Tibetan Buddhism: Indian Buddhists and Their Tibetan Successors.* 2 vols. Boston: Shambhala, 1987.

Srinivasan, Doris M. "The So-called Proto-Śiva Seal from Mohenjo-Dāro: An Iconographic Assessment." *Archives of Asian Art* 29 (1975–76): 47–58.

———. "Vedic Rudra-Śiva." *Journal of the American Oriental Society* 103 (1983): 543–56.

Staal, J. Frits. *The Science of Ritual.* Professor P. D. Gune Memorial Lectures, ser. 1; Post-graduate and Research Department, Series, no. 15. Puṇe: Bhandarkar Oriental Research Institute, 1982.

———. *The Fidelity of Oral Tradition and the Origins of Science.* Mededelingen der Koninklijke Nederlandse Akademie von Wetenschappen, Adf. Letterkunde, NS 49.2. Amsterdam: North Holland, 1986.

———. *Universals: Studies in Indian Logic and Linguistics.* Chicago: University of Chicago Press, 1988.

Taube, Manfred. *Beiträge zur Geschichte der medizinischen Litertur Tibets.* Monumenta Tibetica Historica, no. 1.1. Sankt Augustin: VGH Wissenschaftsverlag, 1981.

Thomas, E. J. *The History of Buddhist Thought.* London: Routledge & Kegan Paul, 1959.

Trenckner, V., et al., eds. *A Critical Pāli Dictitionary.* Vols. 1–. Copenhagen: Ejnar Munksgaard, 1924–.

Ui, Hakuju et al., eds. *A Complete Catalogue of the Tibetan Buddhist Canons.* Sandai, Japan: Tôhoku Imperial University, 1934.

Unschuld, Paul U. *Medicine in China: A History of Ideas.* Berkeley: University of California Press, 1985.

Vats, Madho Sarup. *Excavations at Harappā: Being an Account of Archaeological Excavations at Harappā Carried Out between the Years 1920–21 and 1933–34.* 2 vols. Delhi: Manager of Publications, Government of India Press, 1940.

Vogel, Claus, "On Bu-ston's View of the Eight Parts of Indian Medicine." *Indo-Iranian Journal* 6 (1962): 290–94.

Warder, A. K. "On the Relationship Between Early Buddhism and Other Contemporary Systems." *Bulletin of the School of Oriental and African Studies* 17 (1956): 43–63.

———. *Indian Buddhism.* Delhi: Motilal Banarsidass, 1970.

Weiss, Mitchell G. "*Caraka Saṃhitā* on the Doctrine of Karma." In *Karma and Rebirth in Classical Indian Traditions*, edited by Wendy D. O'Flaherty, 90–115. Berkeley: University of California Press, 1980.

Wezler, Albrecht. "On the Quadruple Division of the Yogaśāstra, the

Caturvyūhatva of the Cikitsāśāstra and the 'Four Noble Truths' of the Buddha." *Indologica Taurinensia* 12 (1984): 290–337.

Wheeler, Sir R. E. Motimer *Early India and Pakistan to Aśoka*. Rev. ed. New York: Praeger, 1968.

Wijayaratna, Mohan. *Le Moine bouddhiste selon les textes du Theravâda*. Paris: Les Editions du Cerf, 1983.

Winternitz, Moriz. *History of Indian Literature*. Vol 2. Translated by S. Ketkar and H. Kohn. 1927. Reprint. New Delhi: Oriental Books Reprint, 1977.

Witzel, Michael. "On the Origin of the Literary Device of the 'Frame Story' in Old Indian Literature." In *Hinduismus und Buddhismus* (Festschrift für Ulrich Schneider), edited by Henry Falk, 380–413. Freiburg: Hedwig Falk 1987.

Woodward, F. L., and E. M. Hare et al. *Pāli Tripiṭakam Concordance*: *Being a Concordance in Pāli to the Three Baskets of Buddhist Scriptures in Indian Order of Letters*. Vols. 1–3, pt. 6. London: Luzac, 1955–84.

Yuyama, Akira, *Vinaya-Texte*. In Systematische *Übersicht über die Buddhistische Sanskrit-Literature* (*A Systematic Survey of Buddhist Sanskrit Literature*), edited by Heinz Bechert. Pt. 1. Wiesbaden: Franz Steiner Verlag, 1979.

Zimmermann, Francis. *The Jungle and the Aroma of Meats*. Berkeley: University of California Press, 1987 (*La Jungle et le fumet des viandes*. Paris: Gallimard, Le Seuil, 1982).

Zysk, Kenneth G. "Review of Jean Filliozat, *Yogaśataka*." *Indo-Iranian Journal* 23 (1981): 309–13.

———."Studies in Traditional Indian Medicine in the Pāli Canon: Jīvaka and *Āyurveda*." *Journal of the International Association of Buddhist Studies* 5 (1982): 70–86.

———. *Religious Healing in the Veda, with Translations and Annotations of Medical Hymns from the* Ṛgveda *and the* Atharvaveda *and Renderings from the Corresponding Ritual Texts*. Transactions of the American Philosophical Society, no. 75.7. Philadelphia: American Philosophical Society, 1985.

———. "Towards the Notion of Health in the Vedic Phase of Indian Medicine." *Zeitschrift der Deutschen Morgenländischen Gesellschaft* 135 (1985): 312–18.

———. "The Evolution of Anatomical Knowledge in Ancient India, with Special Reference to Cross-Cultural Influences." *Journal of the American Oriental Society* 106 (1986): 687–705.

———. "Mantra in *Āyurveda*: A Study of Magico-Religious Speech in Ancient Indian Medicine." In *Understanding Mantra*, edited by Harvey Alper, 123–43. Albany: State University of New York Press, 1989.

———. "Review of G. J. Meulenbeld and Dominik Wujastyk, eds., *Studies on Indian Medical History*." *Indo-Iranian Journal* 32 (1989): 322–27.

———. "Review of Francis Zimmermann, *The Jungle and the Aṛoma of Meats*." *Indo-Iranian Journal*. Forthcoming.

Index of Sanskrit and Pāli Words

Sanskrit entries are followed by those in Pāli. Both indexes are in Sanskrit alphabetical order.

Sanskrit

Pāli

General Index

Abhaya (son of Bimbisāra), 53
Acupuncture, 48, 54, 65–66, 151 n. 10
Afflictions (external), types of, 15
Aggivesāyaṇa, 27
Agniveśa, 4, 27, 33
Ājīvikas, 27, 102, 105
Ākāsagotta, 43, 84, 114–15
Albīrūnī, 36
Alexander of Macedon, 65
Amrapāli, 53
Ānanda, 85, 126
Anatomical knowledge: aquisition of, 34–37; evolution of, 35
Anatomy, 16, 34–37
Arundhatī, 18–19
Āryans, 13
Asceticism, in Harappan culture, 12
Ascetics, 110, 143 n. 57, 145 n. 7, 163 n. 100; Buddhist, 5, 34–35; Heterodox, 5, 27, 117; physicians, 27–33, 37. *See also* *Śramaṇas*
Aśoka (king), 44, 118
Aṣṭāṅgahṛdaya Saṃhitā, 48, 64
Aṣṭavaidya, 64
Aśvins, 4, 14, 22–23, 25, 139 n. 5, 140 n. 15
Atharvaveda, 14, 19, 25–26, 32, 47, 103, 137 n. 9, 163 n. 96
Ātreya, 4, 46, 54–55, 57, 147–48 n. 41
Ātreya, Bhikṣu, 55
Ātri Piṅgala, 55
Āyurveda, 3, 20, 37, 47, 64, 68, 71, 104, 117–19; characteristics of, 3. *See also* Indian medicine

Bagchi, B. C., 67
Bahiṣpavamāna (Strotra), 22, 139 nn. 4–5
Basham, A. L., 27
Bechert, Heinz, 141 n. 21
Belaṭṭahasīsa, 85
Bhadrakāpya, 31
Bhagavatī Sūtra. *See* *Vyākhyāprajñapti*
Bhaiṣajyaguru (Teacher of Healing), 62, 68; as healing Buddha, 62, 144 n. 2, 152 n. 25
Bhaiṣajyagurusūtra, 62
Bhaiṣajyarājan (Royal Physician), 55, 62, 144 n. 2
Bhaiṣajyasamudgata (Supreme Healer), 62
Bhaiṣajyavastu, 52, 150 n. 7
Bhela, 4
Bhela Saṃhitā, 4, 67, 71, 84
Bhesajjakkhandhaka, 52, 107, 121
*Bhikkhunī*s (nuns), 39–41